Mathematics in Mind

The monographs and occasional textbooks published in this series tap directly into the kinds of themes, research findings, and general professional activities of the Fields Cognitive Science Network, which brings together mathematicians, philosophers, and cognitive scientists to explore the question of the nature of mathematics and how it is learned from various interdisciplinary angles.

This series covers the following complementary themes and conceptualizations:

Connections between mathematical modeling and artificial intelligence research; math cognition and symbolism, annotation, and other semiotic processes; and mathematical discovery and cultural processes, including technological systems that guide the thrust of cognitive and social evolution

Mathematics, cognition, and computer science, focusing on the nature of logic and rules in artificial and mental systems

The historical context of any topic that involves how mathematical thinking emerged, focusing on archeological and philological evidence

Other thematic areas that have implications for the study of math and mind, including ideas from disciplines such as philosophy and linguistics

The question of the nature of mathematics is actually an empirical question that can best be investigated with various disciplinary tools, involving diverse types of hypotheses, testing procedures, and derived theoretical interpretations. This series aims to address questions of mathematics as a unique type of human conceptual system versus sharing neural systems with other faculties, whether it is a series-specific trait or exists in some other form in other species, what structures (if any) are shared by mathematics language, and more.

Data and new results related to such questions are being collected and published in various peer-reviewed academic journals. Among other things, data and results have profound implications for the teaching and learning of mathematics. The objective is based on the premise that mathematics, like language, is inherently interpretive and explorative at once. In this sense, the inherent goal is a hermeneutical one, attempting to explore and understand a phenomenon—mathematics—from as many scientific and humanistic angles as possible.

More information about this series at http://www.springer.com/series/15543

Jacek Woźny

How We Understand Mathematics

Conceptual Integration in the Language of Mathematical Description

Jacek Woźny
Institute of English Studies
University of Wrocław
Otmuchów, Poland

ISSN 2522-5405 ISSN 2522-5413 (electronic)
Mathematics in Mind
ISBN 978-3-030-08513-1 ISBN 978-3-319-77688-0 (eBook)
https://doi.org/10.1007/978-3-319-77688-0

Mathematics Subject Classification (2010): 00-XX, 00-02, 00A30, 00A35, 97-XX, 97-02, 97C30, 97C70, 97D20, 97E40, 97E60, 97H20

Softcover re-print of the Hardcover 1st edition 2018

Printed on acid-free paper

This Springer imprint is published by the registered company Springer International Publishing AG part of Springer Nature.
The registered company address is: Gewerbestrasse 11, 6330 Cham, Switzerland

Acknowledgments

I express my most sincere gratitude and appreciation to Professors Mark Turner (CWRU) and Francis Steen (UCLA) for their support and advice which made this book possible.

Jacek Woźny

Contents

Chapter 1
Introduction

1.1 The Effectiveness of Mathematics, Conceptual Integration, and Small Spatial Stories

On July 20, 1969, the lunar module of Apollo 11 landed on the moon. The trajectory of this historic space flight has been calculated by hand by a group of the so-called human computers.[1] It is just an example of the effectiveness of mathematics in modeling (and changing) the world around us. Mathematics continues to be productively applied in engineering, medicine, chemistry, biology, physics, social sciences, communication, and computer science, to name but a few. As Hohol (2011: 143) points out, this fact is often treated by philosophers as an argument for mathematical realism of the Platonian or Aristotelian variety. It is from this perspective that Quine-Putnam's "indispensability argument," Heller's "hypothesis of the mathematical rationality of the world," and Tegmark's "mathematical universe hypothesis" have been discussed. Eugene Wigner, a physicist, often quoted in this context, finished his paper titled *The Unreasonable Effectiveness of Mathematics in the Natural Sciences* in the following way:

> The miracle of the appropriateness of the language of mathematics for the formulation of the laws of physics is a wonderful gift which we neither understand nor deserve. We should be grateful for it and hope that it will remain valid in future research and that it will extend, for better or for worse, to our

[1] Including an African-American NASA mathematician, Katherine G. Johnson, recently made famous by the highly acclaimed film *Hidden Figures* (2016).

J. Woźny, *How We Understand Mathematics*, Mathematics in Mind,
https://doi.org/10.1007/978-3-319-77688-0_1

> pleasure, even though perhaps also to our bafflement, to wide branches of learning. (1960: 14)

James C. Alexander, a professor of mathematics, also sees the "unreasonable effectiveness" of mathematics as of a mystery but offers the following explanation for it:

> It is a mystery to be explored that mathematics, in one sense a formal game based on a sparse foundation, does not become barren, but is ever more fecund. I posit [...] that mathematics incorporates blending (and other cognitive processes) into its formal structure as a manifestation of human creativity melding into the disciplinary culture, and that features of blending, in particular emergent structure, are vital for the fecundity. (Alexander 2011: 3)

I agree with the above solution to the puzzle and have no doubt that it deserves further study. The subject of this book, further explained in the next section, is to prove that conceptual blending (integration), paired with "the human ability for story" (Turner 2005: 4), accounts for the effectiveness of mathematics. One could add, paraphrasing Wigner, that those two correlated mental features of the human mind make the effectiveness of mathematics reasonable. The conceptual blending theory mentioned by James Alexander in the above quotation is thus introduced by Evans and Green (2006):

> Blending Theory was originally developed in order to account for linguistic structure and for the role of language in meaning construction, particularly 'creative' aspects of meaning construction like novel metaphors, counterfactuals and so on. However, recent research carried out by a large international community of academics with an interest in Blending Theory has given rise to the view that conceptual blending is central to human thought and imagination, and that evidence for this can be found not only in human language, but also in a wide range of other areas of human activity, such as art, religious thought and practice, and scientific endeavour, to name but a few. Blending Theory has been applied by researchers to phenomena from disciplines as diverse as literary studies, mathematics, music theory, religious studies, the study of the occult, linguistics, cognitive psychology, social psychology, anthropology, computer science and genetics. (401)

Over the last two decades, the importance of conceptual blending and other mental processes in mathematics has been extensively studied by, among others, Lakoff and Núñez (2000), Fauconnier and Turner (2002), Turner (2005), Núñez (2006), Alexander (2011), Turner (2012), and Danesi (2016). Let us just quote two little fragments, starting with the groundbreaking *Where Mathematics Comes*

From: How the Embodied Mind Brings Mathematics Into Being by George Lakoff and Raphael Nunez.

> Blends, metaphorical and nonmetaphorical, occur throughout mathematics. Many of the most important ideas in mathematics are metaphorical conceptual blends (2000: 48)

Mark Turner adds the concept of "small spatial story" as a vital component of conceptual blending in mathematics:

> Our advanced abilities for mathematics are based in part on our prior cognitive ability for story [...] - understanding the world and our agency in it through certain kinds of human-scale conceptual organizations involving agents and actions in space. Another basic human cognitive operation that makes it possible for us to invent mathematical concepts [...] is "conceptual integration," also called "blending." Story and blending work as a team." (2005: 4)

Considering the already existing, impressive body of the literature on the subject of cognitive exploration of mathematics, we might question the point of adding yet another text to it; however, we have to bear in mind that mathematics is a vast discipline that has been evolving over millennia—there are still vast "here be dragons" areas on the map. All of the existing studies so far are case studies—usually focusing on a few selected mathematical concepts. For example, the foundational text by Lakoff and Nunez (2000) covers set theory, algebra, and various selected topics like infinity, complex numbers, and Euler's equation. However, its coverage of algebra is about 10 pages long (110–119), and this is certainly not enough for one of the most important branches of mathematics. The other sources I mentioned above (Fauconnier and Turner 2002; Turner 2005; Núñez 2006; Alexander 2011; Turner 2012; Danesi 2016) are equally selective in their choice of mathematical topics. And this is why a more comprehensive approach, further described in the next section, is called for.

1.2 The Point and Method of the Book

I will prove that the construction of meaning in mathematics relies on the iterative use of basic mental operations of *story* and *blending* and demonstrate exactly how those two mental operations are responsible for the effectiveness and fecundity of mathematics. It will be done by analyzing the language, the primary notions, axioms, definitions, and proof in Herstein's (1975) excellent *Topics in Algebra*—a classic handbook[2] addressed to "the most gifted sophomores in mathematics at Cornell" (8). Possible further effects of this study are making mathematics more accessible (easier to teach and learn) and perhaps demystifying mathematics as a product of the human mind rather than some eternal Platonic ideal.[3]

The research is systematic in two ways. Firstly, it covers all crucial areas of modern algebra, focusing on the fundamental notions such as set and element, mapping, group, binary operation, homomorphism, ring, and vector space. Secondly, it avoids what Stockwell (2002: 5) calls "a trivial way of doing cognitive poetics"—treating a literary (mathematical in our case) text only as a source of raw data to apply some acumen of cognitive psychology and cognitive linguistics. I don't "set aside impressionistic reading and imprecise intuition" (ibid.). The book's scrutiny of mathematical narrative is not limited to just spotting the mental patterns mentioned above but goes further to demonstrate how those universal patterns of "the way we think" influence our understanding of mathematics—the construction of mathematical meaning.

1.3 Who Is the Book Addressed To

The book is addressed to cognitive scientists, cognitive linguists, mathematicians, teachers of mathematics, and anybody interested in explaining the question of how mathematics works and why it works so well in modeling (what we perceive as) the world around us. I could not agree more with Rafael Nunez when he postulates that

---

[2] Undergraduate modern algebra courses are sometimes referred to as "Herstein-level courses."

[3] The philosophical reflection on the ontological status of mathematical entities is beyond the scope of this book, but let us just point out that Platonic realism seems to prevail in this respect among mathematicians, Herstein included. The famous Swiss mathematician and philosopher, Paul Bernays (1935: 5), after analyzing the foundational contributions of Dedekind, Cantor, Frege, Poincare, and Hilbert, concluded, 40 years before the first edition of Herstein's *Topics in Algebra*, that "Platonism reigns today in mathematics".

"mathematics education should demystify truth, proof, definitions, and formalisms" and that "new generations of mathematics teachers, not only should have a good background in education, history, and philosophy, but they should also have some knowledge of cognitive science."[4] Although our focus is academic-level mathematics, I have been trying not to befuddle the reader with too many advanced level formulas. The book, I very much hope, should be easy to follow by someone with no mathematical or cognitive science grounding. And the next chapter, in which the basic concepts are explained, is designed for that very purpose.

1.4 The Organization of the Book

After introducing our main research tools (basic human cognitive abilities) and presenting an overview of our research area (modern algebra) in the next chapter, we will follow the order of a typical university-level algebra course (in our case, Herstein 1975). We will start with analyzing the set theory and mappings (Chaps. 3 and 4, respectively)—considered to be the foundation of the whole edifice of modern mathematics—and continue along the path of increasing complexity to groups (Chap. 5), rings, fields, and vector spaces (Chap. 6). On each of those stages, we will take a close look at the primary concepts, axioms, definitions, and proof to see the telltale traces of the basic human cognitive patterns of story and conceptual blending.

[4] http://www.cogsci.ucsd.edu/~nunez/web/PME24_Plenary.pdf, accessed 12.12.2016.

Chapter 2
The Theoretical Framework and the Subject of Study

2.1 Overview

The following sections will introduce the tools of study and the subject to be studied—mental operations of story and conceptual blending and modern algebra.

2.2 Language, Cognition, and Conceptual Integration

2.2.1 Cognitive Linguistics

Cognitive linguistics is a relatively modern discipline based on the assumption that language reflects patterns of human thought, perception, motor system, and bodily interactions with the environment. As Eve Sweetser concisely puts it, "Linguistic system is inextricably interwoven with the rest of our physical and cognitive selves" (1990: 6). Evans and Green (2006) describe the origin of cognitive linguistics in the following way:

> Cognitive linguistics [...] originally emerged in the early 1970s out of dissatisfaction with formal approaches to language. Cognitive linguistics is also firmly rooted in the emergence of modern cognitive science in the 1960s and 1970s, particularly in work relating to human categorisation, and in earlier traditions such as Gestalt psychology. Early research was dominated in the 1970s and 1980s by a relatively small number of scholars. By the early 1990s, there was a growing proliferation of research in this area, and of researchers who identified themselves as 'cognitive linguists'. In 1989/90, the International

J. Woźny, *How We Understand Mathematics*, Mathematics in Mind,
https://doi.org/10.1007/978-3-319-77688-0_2

> Cognitive Linguistics Society was established, together with the journal Cognitive Linguistics. In the words of the eminent cognitive linguist Ronald Langacker (1991: xv), this 'marked the birth of cognitive linguistics as a broadly grounded, self conscious intellectual movement'. (3)

One of the reasons for the described above, rapid expansion of the discipline was the fact that language, fascinating as it is, does no longer have to be studied for its own sake.

> An important reason behind why cognitive linguists study language stems from the assumption that language reflects patterns of thought. Therefore, to study language from this perspective is to study patterns of conceptualisation. Language offers a window into cognitive function, providing insights into the nature, structure and organisation of thoughts and ideas. The most important way in which cognitive linguistics differs from other approaches to the study of language, then, is that language is assumed to reflect certain fundamental properties and design features of the human mind. (Evans and Green 2006: 5)

By studying linguistic patterns within this theoretical frame, researchers gain access to the universal patterns of human thought—to "the way we think." And "the way we think," not accidentally, is how Mark Turner and Gilles Fauconnier (2002) chose to entitle[1] their groundbreaking book on conceptual integration theory, which is now part of the cannon of cognitive linguistics and also the subject of the following section.

2.2.2 *Conceptual Integration (Blending) Theory: The Basic Architecture*

Three theories feature prominently in cognitive semantics: cognitive metaphor theory,[2] mental spaces theory,[3] and conceptual integration theory,[4] the latter related to the previous two and often described as an extension of them.

---

[1] The full title is *The Way We Think: Conceptual Blending and the Mind's Hidden Complexities*.

[2] cf. Lakoff and Johnson (1980), Lakoff and Turner (1989), Lakoff (1993), Gibbs and Steen (1999), Lakoff and Johnson (1999)

[3] cf. Fauconnier (1994), Fauconnier (1997), Fauconnier and Sweetser (1996)

[4] cf. Fauconnier and Turner (1998), Coulson and Oakley (2000), Fauconnier and Turner (2002)

> Blending Theory is most closely related to Mental Spaces Theory, and some cognitive semanticists explicitly refer to it as an extension of this approach. This is due to its central concern with dynamic aspects of meaning construction and its dependence upon mental spaces and mental space construction as part of its architecture. However, Blending Theory is a distinct theory that has been developed to account for phenomena that Mental Spaces Theory and Conceptual Metaphor Theory cannot adequately account for. Moreover, Blending Theory adds significant theoretical sophistication of its own. The crucial insight of Blending Theory is that meaning construction typically involves integration of structure that gives rise to more than the sum of its parts. Blending theorists argue that this process of conceptual integration or blending is a general and basic cognitive operation which is central to the way we think. (Evans and Green 2006: 400)

Mark Turner (2014) begins his book, titled *The Origin Of Ideas: Blending, Creativity, And The Human Spark*, with the following statement:

> The human contribution to the miracle of life around us is obvious: We hit upon new ideas, on the fly, all the time, and we have been performing this magic for, at the very least, 50,000 years. We did not make galaxies. We did not make life. We did not make viruses, the sun, DNA, or the chemical bond. But we do make new ideas—lots and lots of them. [...] Each of us is born with this spark for creating and understanding new ideas. But where exactly do new ideas come from? The claim of this book is that the human spark comes from our advanced ability to blend ideas to make new ideas. Blending is the origin of ideas. (1)

Blending then is the way we construct meaning and create new ideas, but what is it exactly? James Alexander (2011) begins his explanation of conceptual blending in the following way:

> Blending is a common but sophisticated and subtle mode of human thought, somewhat, but not exactly, analogous to analogy, with its own set of constitutive principles, explicated, for example, in Fauconnier and Turner's book *The Way We Think: Conceptual Blending and the Mind's Hidden Complexities*. (Alexander 2011: 2)

Blending, as we learn, is "somewhat, but not exactly, analogous to analogy"— does not sound very precise, does it? But James Alexander is perfectly right—let us take a closer look at "analogy." In the next section in Table 2.4, we will have an example of "thinking in terms of"— a rather frivolous proof that thinking is (like) a camping trip in the Lake District. The left column is "somewhat analogous" to the right column. For example, solving a problem is analogous to cracking a hard-boiled egg (or a walnut if it is a tough one). And we can easily see this analogy (or metaphor). In both cases (a problem, a walnut), prolonged effort, applying pressure, is involved. In both cases we are trying to get inside, to uncover something that is hidden, and—if successful—we are rewarded. This analogy, or metaphor, can be described as a mapping from the domain of cracking walnuts to the domain of solving complex theoretical problems (say, solving a differential equation). And we are now "somewhat but not exactly" there. Let me remind the reader—we are trying to explain what conceptual blending is. So far we have established a set of analogies:

the walnut cracker (person) - the mathematician
walnut shell - the mathematical difficulty
physical effort, pressure - mental effort
peeling the walnut - constructing the solution
the content of the shell - the satisfaction of solving the equation
the nutcracker (tool) - the Calculus

In conceptual metaphor theory, the above would be called *the metaphorical mapping*. But human imagination is capable of more than that, more than just mapping the existing elements. For example, we can now imagine a person who uses advanced mathematics to find the best methods of cracking the walnut shell. This brilliant mathematician/walnut enthusiast 1 day invents a perfect nutcracking machine, sells the patent to Kellogg's, becomes immensely rich, gets bored with life and drinks herself to death, etc. We are capable of integrating, merging, compressing the input elements (the walnut cracker, the mathematician), importing new elements (Kellog's, patent office, money, drinking habit), and then imaginatively running the story, inventing a whole new scenario. And after that, we may look back at the mathematician and the walnut cracker in the new light— in the blending theory, it is called "projecting back from the blended space to the input spaces." The process of blending is also referred to as building a conceptual integration network. This is how Fauconnier and Turner (2002), the creators of conceptual blending theory, describe it:

> Building an integration network involves setting up mental spaces, locating shared structures, projecting backwards to inputs, recruiting new structure to the inputs or the blend, and running various operations in the blend itself. (44)

The four mental spaces mentioned above are represented schematically in Fig. 2.1.

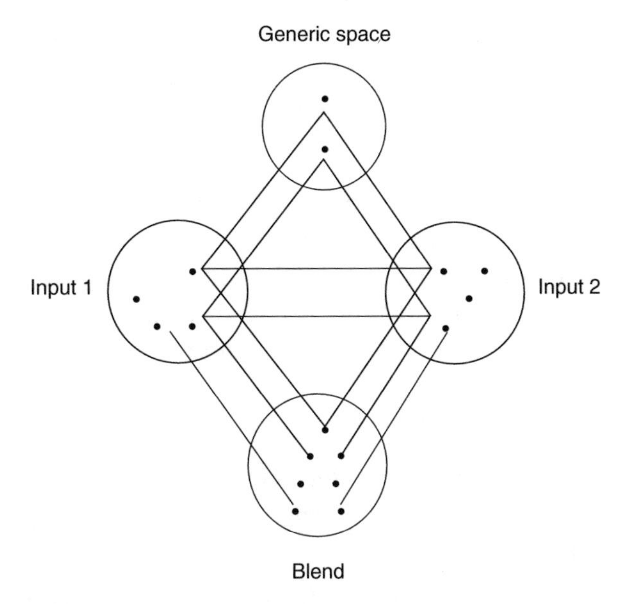

Fig. 2.1 Schematic representation of a conceptual integration network (Evans and Green 2006: 405)

In our "nutcracker example," one of the input spaces is the small spatial story of a person trying to crack a nut, and the other represents the mathematician trying to solve an equation. The generic space contains the shared features—the analogies between the two—and the blend (or "blended") space is where the action of compressing the two stories takes place. The operations taking place in the blend space are the already exemplified compression, completion, and elaboration (imagining the new scenario, also called "running the blend"). The lines represent selective mappings between the spaces.

In the following two sections, we will discuss the criticism of the conceptual integration theory as well as Fauconnier and Turner's (2002) reaction to the most salient critical points in the form of "the constitutive and governing principles."

2.2.3 The Criticism of the Conceptual Integration Theory

Despite the vigorous proliferation of CIT[5]-based research in fields as diverse as linguistics, ethnography, literary studies, and mathematics—or perhaps as a result of it—Fauconnier and Turner's theory has

[5] Conceptual integration theory (aka conceptual blending theory). Stadelmann (ibid.) uses the abbreviation MSCI (Mental Spaces & Conceptual Integration)

been a subject of a lively critical debate over the last two decades. Stadelmann (2012: 28–39) provides a long list of critical points made against conceptual integration theory:

1. Lack of terminological clarity
2. Using only post hoc evidence
3. Neglecting social, material, and historical dimension of cognition
4. Doubtful psychological reality of the generic space
5. The theoretical inconsistency of the simplex network
6. Lack of clear delineation and connection of backstage and onstage cognition
7. Unconstrained character (unfalsifiability)

Ad. 1 (lack of terminological clarity)
Stadelmann points out the definitional fuzziness of the basic ingredients of CIT:

> What exactly are Mental Spaces? What is it that they contain? Why are they 'spatial' in nature? Where are their boundaries? What is 'mental' about them? Here as in other areas, Fauconnier & Turner provide little information, and delineating and determining the content of Mental Spaces in a univocal manner is virtually impossible. (ibid.: 32)

According to Fauconnier (1997:11), mental spaces are "partial structures that proliferate when we think and talk, allowing a fine-grained partitioning of our discourse and knowledge structures." And elsewhere (1985: 8), he defines them as "partial assemblies constructed as we think and talk, for purposes of local understanding and action [...] structured by frames and cognitive models." It is difficult to disagree with Stadelmann—those "definitions" are vague and incomplete. On the other hand—as we will see especially in Chaps. 3 and 4—algebra, usually considered a model of scientific rigor, is based on the so-called primitive notions of set, element, and ordered pair, which are never defined. And, just like mathematics, despite this definitional fuzziness, CIT continues to prove its applicability in many and diverse research areas.

Ad. 2 (using only post-hoc evidence)
The analyses of CIT are based on the retrospective decomposition of a finished product of a set of mental operations—a detective's reconstruction of events from the evidence found at the crime scene. As Stadelmann puts it:

> Like many other cognitive semantic theories – most prominently Conceptual Metaphor Theory (Lakoff, 1986; Lakoff & Johnson, 1980) – MSCI[6] has been accused of delivering post hoc analyses only, meaning that it is ostensibly unable to account for actual online meaning construction (Gibbs, 2000; Harder, 2003; Hougaard, 2004, 2005). Starting with the 'product' of blending and then working backwards rather than following meaning construction ad hoc may lead to the data being tailored to fit the theory rather than the theory being derived from the data. It further leads to a failure in tracking the process of meaning-making as it unfolds in actual on-line cognising. However, advocates (e.g. Coulson & Oakley, 2000; Rohrer, 2005; Talmy, 2000) of the post-hoc approach argue that products constitute the only data currently available to researchers, and that it is impossible to track the psychological steps taken in any particularly accurate manner. (2012: 29)

The advocates of CIT (in fact, Vera Stadelmann is in this group too) certainly have a point here. Linguistic research is based on language—a product (post hoc evidence) of cognition. However, Stadelmann claims that accounting for social (interactional) aspect of meaning construction can free us from the post hoc reconstruction trap.

> Yet this might only be true for approaches focusing exclusively on the individual mind. When considering the interactional dimension of meaning-making, too, as interactional approaches to Cognitive Semantics have done (most notably Hougaard, 2004, 2005), evidence for step-by-step construction, essentially the processes of joint meaning-making over a number of turns, might be gathered. (ibid.)

The "interactional dimension" of meaning is featured also in the next point of criticism.

Ad. 3 (neglecting social, material, and historical dimension of cognition)

Conceptual integration networks are often analyzed independently of the communicative situation. If the hearer knows that the speaker has a friend called "Achilles," who has recently gown an impressive blond beard, the meaning construction of the utterance "Achilles is a lion" would certainly be influenced by it, especially if the utterance was accompanied by a gesture of pointing toward the face (Stadelmann 2012: 30).

> The role of context in which the individual phenomenon is embedded is largely neglected [...]; the historical, social and material dimension of cognition

[6] Mental spaces and conceptual integration

> and its influence on meaning construction is similarly ignored (see also Harder, 2007; Sinha, 1999). Yet the 'content' of blending scenarios largely depends on the situation in which they are embedded, as has already been pointed out above. The emergent properties of a blend will subsequently differ depending on conversational salience, genre and situational relevance. (ibid.: 31)

Can CIT account for (historical, material, situational) context-dependent multiple ways of meaning construction? It certainly can, even if some of the analyses provided by Fauconnier and Turner (1998, 2002) do not. In the "Achilles is a lion" example above, the mapping would be different (facial hair as a salient element rather than courage and physical strength), but it would still fit the theoretical frame. Multiple context-dependent ways of meaning construction also occur in mathematics, which may be surprising as the latter is considered to be a paragon of scientific rigor. In Chap. 4, for example, we will see that depending on the context (on the "ostensive clue"), the mapping can be understood as a "matchmaker," a "carrier," or a "hiker." The next point of criticism is closely connected with points 2 and 3.

Ad. 4 (doubtful psychological reality of the generic space)
As we learned in the previous section, the generic space contains the shared structures of the two inputs. Let us have an example. Fauconnier and Turner (2002: 39) analyze the following riddle:

> A Buddhist Monk begins at dawn one day walking up a mountain, reaches the top at sunset, meditates at the top for several days until one dawn when he begins to walk back to the foot of the mountain, which he reaches at sunset. Make no assumptions about his starting or stopping or about his pace during the trips. Riddle: Is there a place on the path that the monk occupies at the same hour of the day on the two separate journeys? (Koestler 1964)

According to the authors, solving of the above requires a construction of a conceptual integration network. Input 1 contains the monk moving up, and in input 2 the monk is moving down. The blended space contains two monks, one going down and one going up, who at some point meet. The generic space is defined in the following way:

> Generic Space. A generic mental space maps onto each of the inputs and contains what the inputs have in common: a moving individual and his position, a path linking foot and summit of the mountain, a day of travel, and motion in an unspecified direction.
>
> (Fauconnier and Turner 2002: 41)

The critics of CIT point out that the generic space is a content-independent abstraction of the two inputs and a result of a post hoc analysis, existing in the mind of the linguist and not accessible to the participant of the communicative act.

> The question as to whether generic spaces are psychologically real and necessary for the faithful analysis of blends is suitable to be raised [...]. In this regard, Hougaard (2004) points out that the generic structure proposed only adds abstracts from the input spaces to the network rather than further semantics as well. Consequently, generic spaces are only required in the analyses of decontextualised, isolated examples that are not embedded in local contexts. In these cases, it is only the post-hoc constructed tertium comparationis structure that licenses blends, whereas in contextualised data it is local contexts that sanction, implement relevant blending operations, and guide the structure emerging from the blending process. (cf. Brandt & Brandt, 2005) (Stadelmann 2012: 29)

The lack of psychological reality is featured also in the next point of criticism, which is connected with one of the four basic types of conceptual blending networks.

Ad 5. (The theoretical inconsistency of the simplex network)
The typology of conceptual integration networks is provided in the next section, where we will find that in the so-called simplex network, only one of the input spaces is structured—"contains a frame." Let us have an example.

> An especially simple kind of integration network is one in which human cultural and biological history has provided an effective frame that applies to certain kind of elements and values, and that frame is in one input space and some of those kinds of elements are in the other input space. A readily available frame of human kinship is *the family*, which includes roles for the father, mother, child and so on. This frame prototypically applies to human beings. Suppose an integration network has one space containing only this frame, and another space containing only two human beings, Paul and Sally. When we conceive of Paul as the father of Sally, we have created a blend in which some of the *family* structure is integrated with the elements Paul and Sally. In the blended space, Paul is the father of Sally. This is a simplex network. (Fauconnier and Turner 2002: 120)

The definition and the example seem pretty straightforward, but if we go back to the first point of criticism (lack of terminological clarity), we will remember that mental spaces were defined as "partial assemblies constructed as we think and talk, for purposes of local understanding and action [...] structured by frames and cognitive models" (Fauconnier 1985: 8). And we have just learned that some mental spaces can be unstructured, and some contain "pure structure," reminiscent of Lewis Carol's "cat without a grin" and "grin without a cat." Fortunately, in our analyses of conceptual blending in mathematics, we will deal mostly with fully blown mental spaces, containing actors moving in space (like in the Buddhist monk riddle above) and manipulating objects. Mark Turner (1996) calls them "small spatial stories." (cf. Sect. 2.2.5).

Ad. 6 (lack of clear delineation and connection of backstage and onstage cognition)

According to CIT (MSCI), most of our thought processes are unconscious; they are the "backstage cognition." Fauconnier and Turner (2002: 321) use the metaphor of brain as a "bubble chamber of mental spaces." And only selected mental spaces are brought to our consciousness. But, according to Stadelmann (2012), the conscious/unconscious duality is never properly dealt with in CIT, and their connection remains unclear. How and according to what criteria are mental spaces brought from the unconscious to the conscious cognition?

> MSCI hopes to shed light on invisible 'backstage' cognition through 'network' analyses. The theory thus draws heavily on established philosophical metaphors of stages and net(work)s and leaves open the question regarding what exactly it is that differentiates 'onstage' from 'backstage'. What is it that selects a given integration network from the many that are "attempted and explored in an individual's backstage cognition" (Fauconnier & Turner 2002:309)? The authors deal with this question briefly, stating that "the nature of consciousness is to give us effects we can act on, and these effects are correlated with the unconscious processes" (ibid:56). In other words, "the moment of tangible, global understanding comes when a network has been elaborated in such a way that it contains a solution that is delivered to consciousness" (ibid:57). But who/what is the agent delivering, and to whom is it being delivered? As consciousness and unconsciousness are not discussed in detail (only approximately a page is dedicated to the matter) and lack reference to philosophical or neurobiological discourse, one is left to wonder whether the notion of 'backstage' cognition in MSCI might after all involve a Homunculus translating the 'backstage' to the 'onstage', "selecting" Mental Spaces and blends from the "bubble chamber of Mental Spaces" (ibid:321) that is our brain. (Stadelmann 2012: 35)

I agree with Vera Stadelmann—the criteria of selecting the bubbles from the bubble chamber of our brain remain undiscovered. The stage lighting of human cognition works in mysterious ways. However, this theoretical gap (and all theories have them) might one day be explained. And then, Vera Stadelmann's homunculus, like Maxwel's demon, will have to retreat to serve as a supernatural explanatory factor elsewhere. Of course, the assorted homunculi and demons will never be made redundant.[7] And speaking of redundancy, let us move on to the last and most important point of criticism of the conceptual integration theory.

Ad. 7 (unconstrained character: unfalsifiability)
Of all critical points, the last one—the unfalsifiability (or untestability)—seems the most grave. Any theory should be constrained in two ways: we should know where it can be applied, and—when the theory is applied—we should be able to tell whether it yields correct results.

> Does conceptual blending occur wantonly? Is everything 'blendable'? Early MSCI research was often accused of being too 'unconstrained', of advocating an 'anything goes' theory (cf. Gibbs, 2000), as it could not provide an adequate explanation for constraints on conceptual blending. (Stadelmann 2012: 22)

Stadelmann sees the connection of "the testability problem" of CIT (MSCI) with the two tenets of cognitive linguistics—the so-called generalization commitment and cognitive (converging evidence) commitment (cf. Lakoff 1990).

> A key goal of Cognitive Linguistics in general and Cognitive Semantics in particular lies in identifying the general principles of human cognition that apply across a wide range of phenomena (cf. Fauconnier, 1999). It thus contrasts with approaches that assume separate facilities for different aspects of cognition, such as a "faculty of language" (cf. Hauser, Chomsky, & Fitch, 2002). This leads to the attempt by Cognitive Linguistics to attain 'powerful generalisations', such as those provided by MSCI. After all, conceptual blending is supposed to capture The (general) Way We Think, encompassing such diverse phenomena as constructions, metaphors, art and mathematics. Although carrying out research as a means of arriving at general conclusions regarding human cognition via the collection of "converging evidence" from a variety of fields is in itself laudable, generalisations also generate numerous complicated predicaments. This includes becoming banal, or rather being

[7] Unless we finally find an answer to "the ultimate question of life, the universe and everything" (Douglass Adams)

> unable to provide enlightening insights into specific phenomena and actual human behaviour [...] (Bache, 2005; Hougaard, 2004). (2012: 30)

Does anything go? Is CIT an unconstrained theory of everything with predictions that are too general, even banal? The answers are in the next section.

2.2.4 The Constitutive and Governing Principles

As a response to the alleged "wantonness" (unconstrained character) of their theory, Fauconnier and Turner introduce the set of "constitutive and governing principles":

> Cognitively modern human beings use conceptual integration to innovate-to create rich and diverse conceptual worlds that give meanings to our lives-worlds with sexual fantasies, grammar, complex numbers, personal identity, redemption, lottery depression. But such a panorama of wildly different human ideas and behaviors raises a question: Does anything go? On the contrary, conceptual integration operates not only according to a clear set of constitutive principles but also according to an interacting set of governing principles. (2002: xvi)

The constitutive principles can be considered a blueprint of a conceptual integration network, which is built with the use of mental spaces, selective projection, and compression. Stadelmann (2012) gives the following concise description of the blueprint:

> On the constitutive layer [...] conceptual blending relies on the setting up of Mental Spaces and the mappings occurring between them by means of selective projection; these mappings yield novel, emergent insights that are not found in the respective inputs through selective projection via vital relations. Compression allows for global insight on a human scale to emerge in the blends, which often unite complex and semantically distant scenarios. (22)

The "vital relations" mentioned above are various types of mappings between and inside mental spaces in conceptual integration networks listed in Table 2.1. The right column shows typical compressions of the mappings between mental spaces in the network. For instance, CHANGE is compressed into UNIQUENESS. Evans and Green (2006) consider the following example: "The ugly duckling has become a swan." Despite the complete change in appearance over time, the swan is considered to be the same unique individual (422).

Table 2.1 The list of vital relations (Evans and Green: 425)

Outer-space vital relation	Inner-space vital relation (compression)
TIME	SCALED TIME SYNCOPATED TIME
SPACE	SCALED SPACE SYNCOPATED SPACE
REPRESENTATION	UNIQUENESS
CHANGE	UNIQUENESS
ROLE-VALE	UNIQUENESS
ANALOGY	IDENTITY CATEGORY
DISANALOGY	CHANGE UNIQUENESS
PART-WHOLE	UNIQUENESS
CAUSE-EFFECT (bundled with TIME and CHANGE)	SCALED TIME UNIQUENESS
CAUSE-EFFECT	PROPERTY

Table 2.2 contains further constraints of the conceptual integration theory—the governing principles (also referred to as the *optimality constraints*). The constitutive principles told us how to build a network, and now governing principles provide further details of the construction and—most of all—the rules of proper maintenance.

The typology of typical integration networks is not typically given as part of the constitutive or governing principles, but it can certainly be classified as one of the (empirically based) constraints of the conceptual integration theory. Table 2.3 lists various types of conceptual integration networks. We have to remember, however, that they are the most frequently occurring rather than the only possible ones:

> The multiple possibilities for compression and decompression, for the topology of mental spaces, the kinds of connections among them, the kinds of projection and emergence, and the richness of the world produce a vast array of possible kinds of integration network. Amid this diversity, four kinds of integration network stand out: simplex, mirror, single-scope and double-scope [...] and, indeed, when we look at the laboratory of Nature, we find very strong evidence that they really exist. (Fauconnier and Turner 2002: 119)

Fauconnier and Turner (2002) summarize the constraints of conceptual blending in the following way:

> The principles of conceptual integration - constitutive and governing - have been discovered through analysis of empirical data in many domains. These

Table 2.2 The list of governing principles (Evans and Green 2006: 433)

Governing principle	Definition
The topology principle	Other things being equal, set up the blend and the inputs so that useful topology in the inputs and their outer-space relations is reflected by inner-space relations in the blend (Fauconnier and Turner 2002: 327)
The pattern completion principle	Other things being equal, complete elements in the blend by using existing integrated patterns as additional inputs. Other things being equal, use a completing frame that has relations that can be compressed versions of the important outer-space vital relations between the inputs (Fauconnier and Turner 2002: 328)
The integration principle	Achieve an integrated blend (Fauconnier and Turner 2002: 328)
The maximization of vital relations principle	Other things being equal, maximize vital relations in the network. In particular, maximize the vital relations in the blended space and reflect them in outer-space vital relations (Fauconnier and Turner 2002: 330)
The web principle	Other things being equal, manipulating the blend as a unit must maintain the web of appropriate connections to the input spaces easily and without additional surveillance of composition (Fauconnier and Turner 2002: 331)
The unpacking principle	Other things being equal, the blend all by itself should prompt for the reconstruction of the entire network (Fauconnier and Turner 2002: 332)
The relevance principle	Other things being equal, an element in the blend should have relevance, including relevance for establishing links to other spaces and for running the blend. Conversely, an outer-space relation between the inputs that is important for the purposes of the network should have a corresponding compression in the blend (Fauconnier and Turner 2002: 333)

Table 2.3 Basic types of integration networks (Evans and Green: 431)

Network	Inputs	Blend
Simplex	Only one input contains a frame	Blend is structured by this frame
Mirror	Both inputs contain the same frame	Blend is structured by the same frame as inputs
Single scope	Both inputs contain distinct frames	Blend is only structured by one of the input frames
Double scope	Both inputs contain distinct frames	Blend is structured by aspects of both input frames

principles, with all their intricacies and technical mechanisms, conspire to achieve the goal

- Achieve Human Scale

with noteworthy subgoals:

- Compress what is diffuse.
- Obtain global insight.
- Strengthen vital relations.
- Come up with a story.
- Go from Many to One. (322)

Mark Turner (2005: 4), already quoted in the Introduction, claims that "story and blending work as a team." In the next section, we will take a closer look at this teammate of conceptual integration—the small spatial stories.

2.2.5 Small Spatial Stories and Image Schemas

A small spatial story has three vital components—actors, space, and objects. Actors move in space and manipulate objects. This is how Mark Turner describes its importance:

> We are very good at thinking in terms of small spatial stories. We are built for it, and we are built to use small spatial stories as inputs to conceptual blends. In small spatial stories, we separate events from objects and think of some of those objects as actors who perform physical and spatial actions. We routinely understand our worlds by constructing a conceptual integration network in which one of the inputs is a small spatial story. (Turner 2005: 6)

Let us focus on "thinking in terms of small spatial stories." What does it mean exactly? And what is "thinking"? The answer to the last question can be as follows: THINKING IS A ROMP IN THE LAKE DISTRICT.[8] And now I shall prove it. The left column in Table 2.4 contains some typical expressions we use to describe thinking and understanding, and in the right column, we will find their "Lake District" interpretation. We have space, objects, and actors who move and manipulate the objects. It is just an example showing that we

[8]The capitalization may seem excessive here, but I am following a convention adopted by George Lakoff and Mark Johnson in their famous *Metaphors We Live By* (1980).

Table 2.4 Thinking in terms of small spatial stories

In the realm of thought	In the Lake District
It's a lofty subject	There are peaks
I am in deep water here	Lakes
We have to dig deep	And valleys
I am in the fog	The weather changes
I am in complete darkness	Night falls
Let's shed some light on it	We use torches
Now I can see it	Daybreaks
Let's stay with the subject for a little longer	We make camp
And now let's move to another topic.	And break camp
Let me chew on this one.	Eat sandwiches
That's a tough one to crack	Boiled eggs and walnuts
Let's move around this topic	Find the right path
Let's not go this way. Don't touch this one	Avoid dangerous places
Can we turn this argument around? Not easy to grasp	Manipulate objects
Try to see it from my perspective	Enjoy the views
One day, we will find the answers; get to the truth of the matter	After a long trek, we finally arrive at the overcrowded car park in Windermere.

think about thinking in terms of small spatial stories of actors/agents manipulating objects in space. So, "thinking in terms of small spatial stories" means mapping the domain of space, objects, and actions into the abstract domain of mental activity.

The readers familiar with cognitive linguistics and cognitive science literature will of course recognize here the elements of CMT (conceptual metaphor theory).

Let us have another example of thinking in terms of small spatial stories. Mark Turner (2005) uses the example of the structure called "caused motion" in which an agent applies force to an object causing it to move along certain trajectory as in "He threw a ball over the fence" (Goldberg 1995). The syntactic structure is NP-VP-NP-PP, where NP is a noun phrase, VP is a verb phrase, and PP is a prepositional phrase. Apart from the canonical examples with moving objects, like the one above, the same structure can be found in sentences like:

(1) They teased him out of his senses.
(2) I will talk you through the procedure.
(3) I read him to sleep. Turner (2005: 13)

Each of the sentences is understood in terms of an agent causing an object to move in a certain direction, yet neither of the three examples involves an actual application of force or moving along a trajectory.

Small spatial stories, like the ones discussed above (actors moving in space and manipulating objects), are often associated with the concept of "image schemas" in cognitive science and cognitive linguistics literature. Mandler and Canovas (2014: 2–9) state simply that image schemas are simple spatial stories which constitute a crucial part of early, preverbal conceptual development of infants and are built of certain primitives such as container, path, move, into, out of, behind, contact, link, location, etc. Mark Turner (1996) provides the following definition:

> Image schemas are skeletal patterns that recur in our sensory and motor experience. Motion along a path, bounded interior, balance, and symmetry are typical image schemas." (9).

Certainly not all small spatial stories appearing in the following chapters as sources of mathematical concepts could be classified as image schemas—some can be quite complex—but it seems that all of them are composed of image-schematic elements[9] such as the ones listed in Table 2.5.

Table 2.5 Partial list of image schemas (Evans and Green 2006: 190)

SPACE	UP-DOWN, FRONT-BACK, LEFT-RIGHT, NEAR-FAR, CENTRE-PERIPHERY, CONTACT, STRAIGHT, VERTICALITY
CONTAINMENT	CONTAINER, IN-OUT, SURFACE, FULL-EMPTY, CONTENT
LOCOMOTION	MOMENTUM, SOURCE-PATH-GOAL
BALANCE	AXIS BALANCE, TWIN-PAN BALANCE, POINT BALANCE, EQUILIBRIUM
FORCE	COMPULSION, BLOCKAGE, COUNTERFORCE, DIVERSION, REMOVAL OF RESTRAINT, ENABLEMENT, ATTRACTION, RESISTANCE
UNITY/ MULTIPLICITY	MERGING, COLLECTION, SPLITTING, ITERATION, PART-WHOLE, COUNT-MASS, LINK(AGE)
IDENTITY	MATCHING, SUPERIMPOSITION
EXISTENCE	REMOVAL, BOUNDED SPACE, CYCLE, OBJECT, PROCESS

[9]To learn more about image schemas see, for example, Johnson (1987), Talmy (1988), Brugman (1998), Sweetser (1990), Mandler (1992), Turner (1996).

In *The Literary Mind* (1996), Mark Turner claims that:

> We use story, projection, and parable to think, beginning at the level of small spatial stories. Yet this level, although fully inventive, is so unproblematic in our experience and so necessary to our existence that it is left out of account as precultural, even though it is the core of culture. (Turner 1996: 15)

And if small spatial stories are the "core of culture," essential to our thought and existence, it should not be surprising we will keep finding them again and again in the narrative of mathematics, in the next chapters.

Before we end this section, let us quote the book which created the field of cognitive exploration of mathematics:

> A great many cognitive mechanisms that are not specifically mathematical are used to characterize mathematical ideas. These include such ordinary cognitive mechanisms as those used for the following ordinary ideas: basic spatial relations, groupings, small quantities, motion, distributions of things in space, changes, bodily orientations, basic manipulations of objects (e.g., rotating and stretching), iterated actions, and so on. (Lakoff and Nunez 2000: 29)

In the above quotation, the authors do not use the terms of "small spatial story" or "image schema," but we can easily see the connection. Neither of the two terms can be found in the following quotation either, from Saunders Mac Lane (1909–2005), professor of mathematics at Harvard and Cornell Universities and the president of American Mathematical Society:

> Mathematics is not the study of intangible Platonic worlds, but of tangible formal systems which have arisen from real human activities. (1986: 470)

But the connection to "small spatial stories" (actors moving in space and manipulating objects) is there again. Mac Lane (qtd. in Lakoff 1987: 354) has constructed the following list of correspondences between "real human activities" and branches of mathematics:

> counting: arithmetic and number theory
> measuring: real numbers, calculus, analysis
> shaping: geometry, topology
> forming (as in architecture): symmetry, group theory
> estimating: probability, measure theory, statistics
> moving: mechanics, calculus, dynamics
> calculating: algebra, numerical analysis
> proving: logic
> puzzling: combinatorics, number theory
> grouping: set theory, combinatorics
> (Mac Lane 1986: 463)

This rudimentary sketch of the conceptual integration theory will have to do for now[10]; we will see it in action in Chaps. 3, 4, 5, and 6. The next section offers a bird's-eye view of our research area—the modern algebra.

2.3 Modern Algebra for Beginners

Can the whole of modern algebra be described in a couple of sentences? Yes it can; it has been designed to be elegantly simple. The story starts with sets (collections of objects) and mappings and proceeds to the concept of a group (a set with a mapping), a ring (a set with two mappings), and vector space (two sets with four mappings altogether). An example of a group are integers with addition, real numbers with addition and multiplication have the structure of a ring (also a field), and vector space can be exemplified by complex numbers.[11] Each new concept is based on the previous ones, and, ultimately, the whole multistory edifice rests on the sparse foundation of sets and mappings. Israel Nathan Herstein begins his classic[12] *Topics in Algebra* handbook in the following way:

> One of the amazing features of twentieth century mathematics has been its recognition of the power of the abstract approach. This has given rise to a large body of new results and problems and has, in fact, led us to open up whole new areas of mathematics whose very existence had not even been suspected. [...] The algebra which has evolved as an outgrowth of all this is not only a subject with an independent life and vigor-it is one of the important current research areas in mathematics-but it also serves as the unifying thread which interlaces almost all of mathematics, geometry, number theory, analysis, topology, and even applied mathematics. (Herstein 1975: 1)

---

[10] To learn more about conceptual blending theory, see, for example, Fauconnier and Turner (2002) and Turner (2014).

[11] More precisely, complex numbers are a vector space over the field of real (or complex) numbers (see Chap. 6 for more details).

[12] cf., for example, the Chicago undergraduate mathematics bibliography, where we can read, "[...] classic text by one of the masters [...] wonderful exposition—clean, chatty but not longwinded, informal—and a very efficient coverage of just the most important topics of undergraduate algebra." (https://www.ocf.berkeley.edu/~abhishek/chicmath.htm, accessed 2017-10-06)

The mathematical "abstract approach" mentioned above, also known as the "axiomatic approach"—the origin of modern algebra—was developed gradually in the nineteenth and the first half of the twentieth century.[13] In fact, axioms were used in mathematics ever since the birth of Euclidean geometry (ca. 300 BC), but there is one crucial difference— Euclid defined the primitives, such as point and straight line (e.g., a point is that which has location but no size), while in modern algebra the primary notions, such as set, element, and ordered pair,[14] remain undefined. So, in the next chapters, when we discuss all the fascinating features of groups, we "will not know what we are talking about" to paraphrase the famous statement by Bertrand Russell.[15] And this is because a group is defined as a set with a mapping, which fulfills the group axioms—any set, a collection of any objects. The shortest description of the "abstract approach" could be primary notions + axioms + definitions + theorems + proof. We learn from Herstein's introduction above that algebra is "the unifying thread which interlaces almost all of mathematics"—so this is where we have to look for the foundations of modern mathematics.

As we mentioned above, although our focus is advanced level algebra, reading this book should not require any prior mathematical training. What follows in Table 2.6 is an informal glossary of terms and symbols used in the following chapters in the chronological order. Typically, such glossaries are added at the end of a book, but I think it would be useful for a reader to have a quick look at the key terms now,

---

[13] According to Nicolas Bourbaki (a collective pseudonym for a famous group of mathematicians), "The axiomatization of algebra was begun by Dedekind and Hilbert, and then vigorously pursued by Steinitz (1910). It was then completed in the years following 1920 by Artin, Nöther and their colleagues at Göttingen (Hasse, Krull, Schreier, van der Waerden). It was presented to the world in complete form by van der Waerden's book (1930)." (http://www.math.hawaii.edu/~lee/algebra/history.html, accessed 2017–10-06)

[14] Herstein (1975) does not define ordered pairs and neither did Frege (1879). Other mathematicians suggested various definitions. For example, Hausdorff (1914: 32) gave the definition of the ordered pair (a, b) as {{a,1}, {b, 2}}, but, as we argue below, this does resolve the problem implicit circularity of the static definition of a mapping.

[15] "Mathematics may be defined as the subject in which we never know what we are talking about." (https://en.wikisource.org/wiki/Mysticism_and_Logic_and_Other_Essays, accessed 2017–10-06)

Table 2.6 Chronological glossary of the key mathematical terms

Term	Description
Set	Any collection of objects, a set of integers but also three bricks in a suitcase. Primary notion (not defined). Typically marked with a capital letter or curly brackets { }. For example, N, the set of natural numbers (positive integers with zero); R, the real numbers, etc.; {1,2,3} means a set of three numbers— 1,2, and 3
Element of a set	Any object in a collection (set) of objects. Primary notion (not defined). Typically marked with a lower case letter and the symbol ∈. For example, a∈A reads "a is an element of (the set) A" or "a is in A." {x∈R | x > 0} means a set of all positive real numbers
Subset of a set	A set whose all elements are in another set. A ⊂ S reads "A is a subset of S," "A is contained in S" or "S contains A."
Equal sets	A = B if A ⊂ B and B ⊂ A. Two sets are one set if... But how can two sets be one set? If it is one set, how did it become two sets? Can one be two? Can two be one? Find the answers in Chap. 3
Union of two sets	A ∪ B reads "union of (sets) A and B" and is a set containing all elements of A and all elements of B and only those elements
Intersection of two sets	A ∩ B reads "intersection of (sets) A and B" and is a set containing all elements that are both in A and in B and only those elements
Empty set	A set with no elements. Marked with Ø. Polish philosopher and mathematician, Stanisław Leśniewski, the creator of mereology, called it a "theoretical monstrum" and "a set of square circles"(1930: 196)
Ordered pair	A set of two elements which are ordered (one element is first, and the other is second). Marked with (,). For example, (a,b) reads "an ordered pair of a and b." primary notion (not defined, cf. Ftnt. 14)
Cartesian product of two sets	A Cartesian product of A and B is the set of all ordered pairs (a,b) where a is in A and b is in B. Marked as A × B
Cartesian square of a set	A Cartesian product A × A, a set of all ordered pairs (a,b) where both a and b are in A. Interestingly, every Cartesian square contains the so-called diagonal subset which is a set of ordered pairs (a,a). A curious concept because (a,a) is a pair of two elements which are one and the same element

(continued)

Table 2.6 (continued)

Term	Description
Mapping	A mapping from A to B is a subset of A × B in which every element of A is paired with an element of B. This is the definition Herstein calls "rigorous" and then adds that he almost never uses it, preferring a different "way of thinking about mapping." In Chap. 4 Herstein's puzzling reluctance will be explained
Composition of mappings	g(f(x))—Two mappings acting one after another: First x is mapped onto f(x), and then f(x) is mapped onto g(f(x))
Group	A set with a binary operation. For example, integers with addition. The binary operation must follow certain rules (group axioms)—Like the existence of the identity element and the inverse element. See Chap. 5 for details (and for the mathematical beauty of finite groups)
Binary operation	For example, addition or multiplication. For a group G, the binary operation is defined as a mapping from the Cartesian square G × G to G which means that for every ordered pair (a,b), there exists in G a "result of the operation c." for example, for integers under addition, the pair (2,2) is paired with 4, which is typically written as 2 + 2 = 4
Identity element	For example, 0 for addition or 1 for multiplication. We add it or multiply by it, and nothing changes. Every group must contain an identity element
Inverse element	For example, −5 is the inverse element for 5 (under addition) because 5 + (−5) = 0. Under multiplication, the inverse of 5 is 1/5 because 5(1/5) = 1 (and 1 is the identity element for multiplication, just as 0 is for addition). Every element in a group must have an inverse element
Associativity	For example, for multiplication (ab)c = a(bc). The binary operation in a group must be associative
Abelian group	A group where the binary operation gives the same result when applied in any order. Integers with addition, for example, are an abelian group because a + b = b + a. The binary operation with this feature is called "commutative"
Subgroup	A subset of a group which is also a group
Coset	A set obtained by "multiplying" every element of a subgroup by one element of a group. If H is a subgroup of G and a is in G, aH is how we mark the left coset, and Ha is used for the right coset. For any subgroup, all cosets have the same number of elements, are either equal or disjoint, and cover the whole group. And, for some of us, this is where the beauty of finite groups lies

(continued)

Table 2.6 (continued)

Term	Description
Homomorphism	A special kind of "structure-preserving" mapping. For example, if G and H are groups, the mapping f from G to H is a homomorphism if f(ab) = f(a)f(b) for every a,b in G. The square function for real numbers with multiplication could serve as an example because $(ab)^2 = a^2b^2$. If a homomorphism is a 1-to-1 mapping, it is called an "isomorphism"
Order of a set	Number of elements in a set. For example, if a set G has 3 elements then o(G) = 3, which reads as "the order of G is 3"
Lagrange's theorem for finite groups	If H is a subgroup of G, then o(H) is a divisor of o(G). For example, if G has 12 elements, any subgroup can have 1,2,3,4, 6, or 12 elements. It can't have, for example, 5 or 11 elements. In Chap. 5 we will try to see why this theorem is considered beautiful
Symmetry group on the set S	In Herstein's handbook denoted by A(S). It is the set of all 1-to-1 mappings of S onto itself. Such mappings are also called "permutations." It is easy to prove that A(S) with composition of mappings is a group. In fact, historically, before the abstract approach became dominant, this is what groups where in mathematics—Sets of permutations
Cayley's theorem	Every finite group G is isomorphic with a subgroup of A(G) which means that every finite group is in fact a set of permutations. This theorem (Cayley 1854) was crucial for the development of the abstract approach in algebra, showing that the group axioms are "meaningful" because they define the already well-known "concrete" groups of permutations
Ring	An abelian group with additional binary operation, which has to be associative and distributive (see below). The set of integers with addition and multiplication is a ring
Distributive laws	a(b + c) = ab + bc and (a + b)c = ac + bc. The two binary operations in a ring must fulfill the distributive laws. We might remember them from the primary school as "multiplying brackets"
Division ring	A ring in which inverse elements exist for both binary operations for every element of the ring (except for zero under multiplication). The set of real numbers with addition and multiplication is a division ring, for example
Commutative ring	A ring where both binary operations are commutative
Field	A commutative division ring. The sets of real or rational numbers are examples of fields

(continued)

Table 2.6 (continued)

Term	Description
Vector space	A vector space involves an abelian group and a field which have to fulfill certain rules. For example, the Cartesian square of the set of real numbers R^2 is a vector space over R. See Chap. 6 for details
Module	A generalization of a vector space in which the field is replaced with a ring. Every vector space is a module but not vice versa
n-dimensional vector space	A vector space is n-dimensional if there are n (linearly independent) vectors from which all the other vectors can be obtained as a result of the binary operations (adding vectors and multiplying vectors by scalars). It can be proven that every n-dimensional vector space over a field F is isomorphic (which practically means "identical") with F^n

to see how the captivating story of modern algebra develops from the primary notions of set and element, before we start to delve deeper into the subject.

We now know what to look for (small spatial stories and conceptual integration networks) and where (the narrative of modern algebra, Herstein 1975). Let the hunt begin. We will start at the beginning of Herstein's handbook, with the set theory, and then continue our analysis of the mathematical narrative in a step-by-step, linear fashion, without jumping ahead.

Chapter 3
Sets

3.1 Overview

We will begin this chapter with examining the primitive (undefined) notions of "a set" and "an element" and then investigate the basic set-theoretical concepts of subsets, equality of sets, the null set, the union, and intersection of sets. In the final section, we will take a closer look at the language of mathematical proof. At every stage of our close reading of the mathematical narrative, we will be looking for the mental patterns like image schemas (e.g., the container image schema), small spatial stories (actors moving in space, manipulating objects), and conceptual integration.

3.2 The Primitive Notions: Set and an Element

Set theory is commonly believed to be the foundation of modern mathematics. Mathematical stories often begin with terminology and primary (non-defined) notions. These are often explained by appeal to our intuition, often with examples from our everyday experience:

J. Woźny, *How We Understand Mathematics*, Mathematics in Mind,
https://doi.org/10.1007/978-3-319-77688-0_3

> We shall not attempt a formal definition of a set nor shall we try to lay the groundwork for an axiomatic theory[1] of sets. Instead we shall take the operational and intuitive approach that a set is some given collection of objects. [...] we can consider a set as a primitive notion which one does not define. (Herstein 1975: 2)

In other words, to understand the notion of a mathematical set, we are to rely on our experience with collections of objects. A list of notation shortcuts follows, for example, "given a set S, we shall use the notation throughout $a \in S$ to read 'a is an element of S'" (2). So the notation is to "read," to be expanded into a sentence containing another intuitive (never defined) notion of "an element of a set." Our imagination and experience with collections of objects are now to help us understand that an element is one of those objects in the collection. But neither the object (element) nor the collection (set) is defined. They remain undefined also in the standard axiomatic set theory, called ZFC. Here we are, on page 2 of the algebra handbook, at the foundation of mathematics (or in the dark cellar of it that we do not dare visit at night, to use a less optimistic metaphor), and we have been prompted to use our imagination and experience twice. Our concept of collections and elements is coded in language, so let us consider a few random examples from the British National Corpus[2] of sentences containing the word "contains" in Table 3.1.

As we will realize in the subsequent sections, only one out of the ten above is a "good" example of a set and elements. It is the only one that would be used as an example of a set in a handbook of algebra and the only one that fits the image schemas and small spatial stories upon which the set theory (clandestinely) relies. It is example 2, and to explain it, we need to go further into the set theory and see how the primary notions of set and element are used to define subsets, equality, the null set, the union, and the intersection of sets.

---

[1] The canonical today, axiomatic set theory called ZFC, does not include the definition of a set either. The set remains a primary, undefined notion there as well. We should also mention that many important mathematical theorems (e.g., the continuum hypothesis, Suslin hypothesis, diamond principle) were proven to be "independent" of ZFC, which means they can neither be proved nor disproved within this framework. Which of course is one of the reasons some mathematicians contest the claim of the fundamental role of the set theory in modern mathematics.

[2] The British National Corpus, version 3 (BNC XML Edition), 2007. Distributed by Bodleian Libraries, University of Oxford, on behalf of the BNC Consortium. URL: http://www.natcorp.ox.ac.uk/, accessed 2017-10-10.

Table 3.1 A random BNC sample of sentences containing the verb "contains"

Your query "contains" returned 4560 hits in 1387 different texts (98,313,429 words [4048 texts]; frequency, 46.38 instances per million words) (displayed in random order)	
1. FEV 804[a]	The retina *contains* only nine light detectors
2. CDH 26	Each pack *contains* colorant, protective cape, and gloves
3. K93 287	The final syllable *contains* a short vowel
4. HU2 919	The egg lecithin used *contains* five different fatty acids
5. BPC 1814	The Tudor stable block *contains* an exhibition about the battle of Quebec
6. HYA 2248	The section on "Drama and story" *contains* a number of brief practical examples
7. B1F 1343	For all the joys that the world *contains*, reckon that love is the goldmine and those other things but gilded
8. CER 278	The positive charge in a nucleus is due to the several protons it *contains*
9. B7H 1469	Cornwall still *contains* a lot of tin
10. BNL 1406	A basic slow-cooker is an earthenware pot, with a lid, set in a metal frame that *contains* a heating element

[a]The alphanumerical code allows to identify the source text and the position of the key word (in our case "contains") in it, in the British National Corpus

3.3 Subsets and Equal Sets

3.3.1 Subsets

We can now escape the "dark, dusty cellar" of primitive, undefined concepts that are the foundation of the set theory (and, as many claim, the whole edifice of modern mathematics) to enter the sunny realm of proper science, which is free from the constraints of imagination, experience, image schemas, small spatial stories, conceptual blending, human conceptual system, natural language, etc. (or so it would seem). And this is because, equipped with the set of primitives, we can now define concepts in a rigorous way (or so it would seem). The next quotation is a definition:

> The set A will be said to be a subset of the set S if every element in A is an element of S, that is, if $a \in A$ implies $a \in S$. We shall write this as $A \subset S$ which may be read 'A is contained in S' (or, S contains A). (Herstein 1975: 2)

As we mentioned above, "an element of" was a primary notion, but now the definition of "contained in" (only for sets, not elements) is built upon it. It is important to notice that the set theory reserves the notion of a subset only for sets, not elements. The relations of "being an element" and "being a part" (subset) are completely separate. For example, number 1 is an element of {1,2,3} but not a subset of it. Conversely, the one-element set of {1} is a subset of {1,2,3} but not an element of it.[3] If we believed the myth of the language of mathematics being "rigorously precise," we would expect this difference between elements and subsets to be clearly marked. The verb "contain" was used only in the definition of a subset not in the description of an element of a set. And yet, a few pages down, we find for example:

"A nonempty set of positive integers always contains a smallest element" (18), or "another natural characteristic of a group G is the number of elements it contains" (28). Of course, this is not to criticize our excellent source text. Mathematics uses natural language, and polysemy is the natural feature of all natural languages.

In the next sections, we will learn that different image schemas (to use the terminology from Chap. 2—different small spatial stories in one of the input spaces of the conceptual integration network) are at the base of the two concepts. But first, let us have a look at the relation of equality. It is probably the most important relation in mathematics, which—for sets—is defined upon the notion of a subset.

3.3.2 Equal Sets

> What is meant by the equality of two sets? For us this will always mean that they contain the same elements, that is, every element which is in one is in the other and vice versa. In terms of the symbol for the containing relation, the two sets A and B are equal, written $A = B$, if both $A \subset B$ and $B \subset A$. (ibid. 2)

It is so simple, so obvious—two sets are identical if all the elements are identical. In other words, two entities are one entity if a certain condition is met. On the other hand, how can two entities be one

[3] A Polish mathematician, Stanisław Leśniewski, was so taken aback by the unintuitive nature of the naive (and ZFC) set theory that he decided to create his own theory, known today as mereology (the study of parts). In his theory, (created c.a. 1914) element and subset are one and the same, which is also an elegant way of avoiding the famous Russell's paradox. (cf. Sect. 3.8).

entity? If it is one, it is not two, is it? There is something strange going on here. We need multiple identical tokens of one object (a set in this case) so that then we can look at them separately, check if they fulfill a given condition (being a subset of one another), and then decide that yes, they are identical, they are one. We have to understand that we are not talking about two objects that have the same features, look the same, or belong to the same category, like two identical golf balls or two kings of spades from two identically looking decks of cards. There are unique objects in the world, like the Eiffel Tower or the Tower Bridge, and (equally) there are unique objects in mathematics. For example, in Chap. 5 we will learn that in every group (say, the integers with addition), the identity element (zero) is unique. And it means that there is only one number zero. And yet we can still write $0 = 0$ or {Eiffel Tower} = {Eiffel Tower}, and it will not be considered a contradiction, which begs the question of what identity and uniqueness are. The answer can be found in the following quotation by the creators of the conceptual blending theory:

> The recognition of identity, sameness, equivalence, A = A, which is taken for granted in form approaches (mathematics, for example, JW), is in fact a spectacular product of complex, imaginative, unconscious work. Identity and opposition, sameness and difference, are apprehensible in consciousness and so have provided a natural beginning place for form approaches. But identity and opposition are finished products provided to consciousness after elaborate work; they are not primitive starting points, cognitively, neurobiologically, or evolutionarily. (Fauconnier and Turner 2002: 6)

Most cells in a human organism die and are replaced with alarming frequency (from a couple of weeks to a couple of months); we also age and change appearance, place of residence, profession, interests, etc. and yet stay the same and unique throughout our lifetime. In our everyday experience, one object, say, a cup of coffee on my desk, always stays that way—one. Even when I take a sip and put it down, I still think of it as the same cup of coffee. It never turns into two cups of coffee so that I can look at them separately and say, hmm, yes, they are identical; however, in our imagination and memory, it is possible and we do it all the time. It is enough that I close my eyes for a second, look at the cup again, and think, yes, it's the same cup. I could even write it down mathematically, cup1 = cup2, where cup1 was the cup I saw a second ago and cup2 is the cup I see now. The level and the temperature of the liquid inside are different; the position on the desk

Table 3.2 Elements of small spatial stories and traces of conceptual integration found in the narrative of the set theory

[1]*Objects*	Elements of sets, numbers, all kinds of objects that can belong to a collection
Actors	Set, an actor who possesses objects and governs property; set operator (the potter, the setter), performs operations on sets, uniting, intersecting and dividing them. Proof—An actor who collects mathematical proofs but sometimes has to dispose of them to clear the path on his way to the QED spot
Actions	Possessing/belonging (often categorized as a state, or a potential to act, perhaps not a prototypical action but of course, like with all linguistic taxonomies, the border between state and action is fuzzy); combining sets, forming them into new ones (uniting), intersecting, dividing, disposing of objects
Image schemas[a]	Containers with discrete and dimensionless, or voluminous objects (partly opened or tightly shut), an empty container (the null/empty set), part/whole, in-out, full-empty compulsion, blockage, removal of restraint, enablement, source-path-goal, object, superimposition
Conceptual blending	The equality symbol "=" always involves a blend (triggers a conceptual integration network). Multiple tokens of an object are compressed into a unique object. Yet, because the projections are bi-directional and the network is maintained (according to the web principle and the unpacking principle, cf. Sec. 2.2.4), the object can be "one and many" at the same time

[a]The connection between small spatial story and image schema is explained in Sect. 2.2.5

may be slightly different but it is still the same, unique object. At the same time, I am aware of the various stages my coffee went through. I am able to project back from the cup I see now to those other stages/ tokens.

And that is how the cup is one and many at the same time. The interpretation from the standpoint of the conceptual integration theory is that there are two mental input spaces in which cup1 and cup2 are connected with identity relation, and in the blend space (cf. Table 2.2 in Sect. 2.2.4—the web principle and the unpacking principle) the two cups are compressed into one—a unique object. Identity is the product of blending. The relevant conceptual integration network is represented schematically in Fig. 3.1.

Our sensory input and memory always provide us with multiple tokens, but in our mind, through the constant process of blending,

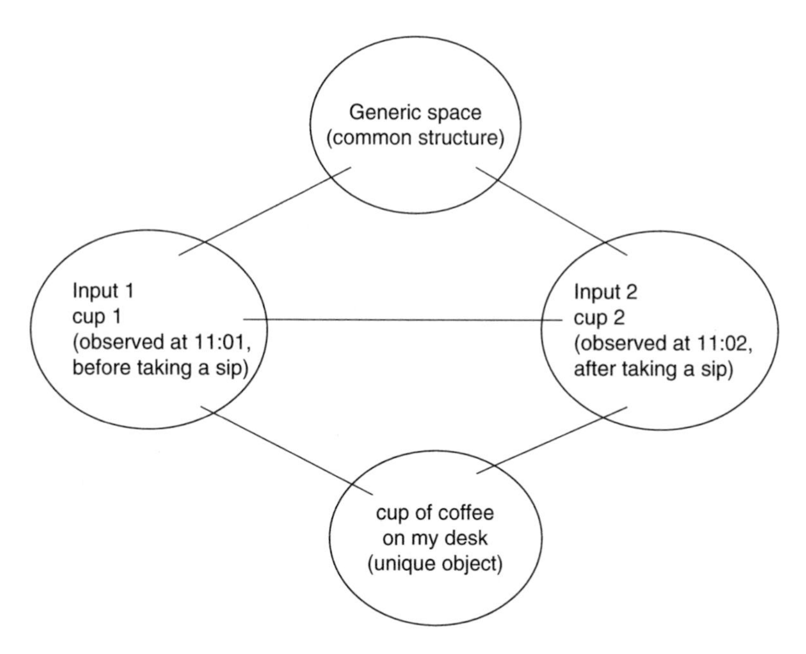

Fig. 3.1 Uniqueness as a product of conceptual blending—a schematic representation

those multiple tokens are compressed into one. Objects are not "externally unique"; uniqueness and identity are unconscious creations of the human mind. We will return to this fascinating subject of uniqueness in Sect. 5.5.

That's all very well but we were just analyzing "pure mathematics"—there is no place in it for conceptual blending, memory, imagination, and sensory perception, all this cognitive stuff. Mathematical entities are eternal and unchanging; they exist in a world separate from the human world, don't they? Well, unfortunately, no. Mathematics, wonderful as it is, is a product of the human mind, and this is just one more example of it. And we have just demonstrated that one of the crucial mathematical notions—equality, expressed with the "=" sign—depends on the human capacity for blending,[4] the same mental process which is responsible for our understanding of the next mathematical notion to discuss, the null (or "empty") set.

[4]We will return to the subject in Chap. 5, when we discuss the uniqueness of the identity element in a group.

3.4 The Null Set

Let us begin with the definition:

> The null set is the set having no elements; it is a subset of every set. We shall often describe that a set S is the null set by saying it is empty. (Herstein 1975: 2)

The notion of an empty set of course only strengthens the claim we have made before that the set-theoretical concepts, and mathematical concepts in general have image schematic origin (cf. Lakoff and Nunez 2000: 29—already quoted above). And, obviously, the empty set points us in the direction of the container image schema. The null set is an empty container. So a set is not just a collection of objects but a collection and a container—the collection in a container. And, as we learn, "it is a subset of every set." Well that's ok, isn't it? All sets also include a container. But apparently they have this empty container inside—as a subset. This certainly goes against our intuition of collections in containers, where the container—the tin, the box, the safe, the display cabinet, the drawer—contain the collection and are not contained in it. Both Frege and Leśniewski strongly criticized the notion, the latter describing the null set as a "theoretical monstrum" and "a class of square circles" (1930: 196).

As we said, the standard set theory is unintuitive and this is just another case in point. And when we say "unintuitive," we are not claiming it is not based on image schemas. On the contrary, it is based on several image schemas, as we tried to demonstrate above, but of course not overtly, which certainly hampers understanding. We have to know which image schema applies to which notion, and this (at least in part) constitutes the base for mathematical meaning and mathematical understanding.

In the following two sections, we will introduce, following Herstein's handbook of course, the two crucial operations on sets: the union and the intersection.

3.5 The Union of Sets

The definition of a union of two sets:

> Given two sets we can combine them to form new sets. The union of the two sets A and B, written as $A \cup B$, is the set $\{x \mid x \in A \text{ or } x \in B\}$. (Herstein 1975: 3)

The sets are "combined to form." The grammar construction used is NP1 verb NP2 to-infinitive NP3, and the narrative structure (the story) associated with it is that an agent (actor) performs an action over objects (two sets) in order to achieve a certain goal (forming a new set). What we have here is a projection of a small spatial story of an actor manipulating objects and forming new objects. The actor performs a purposeful and planned action, according to a template, with predictable result. We may now even imagine a workshop, much like a pottery, where a skilled craftsman combines objects to form new ones. But instead of using bits of wet clay, the craftsman, the set potter (or simply, setter), combines sets to form new sets. We learn further that "for any set A, A ∪ A = A (union of A and A equals A)." Oh, I see, so I have this cup of coffee on my desk (there is always one there), and if I "combine" my cup of coffee with my cup of coffee, I will "form" my cup of coffee, how interesting. We already explained above how the equality sign "=" requires creating two mental tokens (input spaces) of the same object. In this case we have to do it twice in a row. First we need two tokens of A to combine them (to from A again) and then another token of A to compare it to the result of the combining. We already provided many examples of the unintuitive nature of set theory, but this is not one of them. Identity relation is one of the staple connectors, mappings, between mental spaces. And because this mapping is so common and seemingly effortless, it is often considered trivial, simple, and easy.

In the following section, we will focus on another basic set operation—the intersection of two sets.

3.6 The Intersection of Sets

The intersection of sets is defined as follows:

> The intersection of the two sets A and B, written as A ∩ B, is the set {x | x ∈ A and x ∈ B}. The intersection of A and B is thus the set of all elements which are both in A and in B. (Herstein 1975: 3)

It is pictorially represented in Fig. 3.2.

Both the narrative description "all elements which are in" and the image suggest the container schema. However, if we consider an "improper" set X = {set1,set2,set3....X} (cf. Sect. 3.8), no

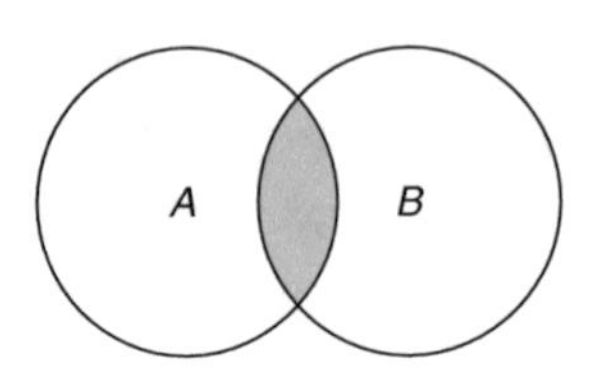

Fig. 3.2 Intersection of sets

image representation could possibly be drawn of a collection which is inside itself as one of the elements—an element that is both inside and outside at the same time. Let us remember also that "being in a set as an element" is a primitive, undefined notion and does not mean the same as "being in a set as a subset," which is defined by the previous primitive.

We can now use our imagination to animate the above image by making the intersection either larger or smaller. And if we move the sets further apart, at one point they will be separated: "Two sets are said to be disjoint if their intersection is empty, that is, is the null set" (ibid.: 4). It is interesting to notice the geometrical differences between the sets and elements. A set can be inside another set or outside it or partially overlap with it—an intermediary state of being partially in and yet "sticking out a little." The elements, however, do not enjoy the same degree of freedom. They can only be either outside or inside a set—no intersection possible. And this difference can be interpreted image schematically, through small spatial stories as we will see in the following section.

3.7 Image Schemas and Small Spatial Stories for Sets and Elements

Image schemas emerge through everyday repeated experience. For the sets and subsets, it might be our experience of a partially open containers, objects allowed to overlap and stick out, like the fruit in a fruit bowl in Fig. 3.3.

For the elements, however, the container is like a box of tiny air gun pellets with a tightly fitted lid in Fig. 3.4.

Figures 3.3 and 3.4 are visualizations of the image schemas, and small spatial stories build on them, which contribute to the meaning construction of sets and elements, respectively. Contribute in the way

Fig. 3.3 Inside, partially inside, outside—an open container containing objects of volume, the schema of sets partially intersecting

Fig. 3.4 Outside or inside, no third option—the closed container containing small, undividable elements, the schema for the elements of a set

described in Sect. 2.2.5. The first small spatial story is "filling a partially open container with objects of volume" (like putting fruit into a fruit bowl). The second story is "filling a container with small point-like objects and then fitting the lid tightly" (the air gun pellets in a jar). The respective small spatial stories become then one of the inputs of conceptual integration networks for sets and elements.

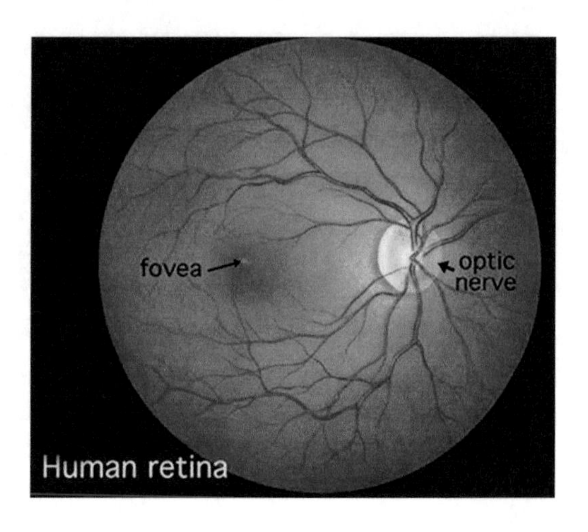

Fig. 3.5 Retina seen through an ophthalmoscope (http://webvision.med.utah.edu/book/part-i-foundations/simple-anatomy-of-the-retina/, accessed 2017-10-10)

And now we can return for a moment to Table 3.1 to explain why only example 2 is the "good one," meaning that it fits the way sets, elements, and subsets should be understood. Let us go through the examples one by one.

1. The retina contains only nine light detectors.

Set theoretically, this example would only make sense if the retina were a set of separate light detectors, and not a continuous, spherical layer of tissue represented in Fig. 3.5. A mathematical set is a collection of separate, undividable, never intersecting objects in a container, and a subset must contain the same elements as the set it is contained in (cf. the definition of subset in Sect. 3.3.1).

2. Each pack contains colorant, protective cape, and gloves.

A perfect example, a set consisting of separate, solid, not-intersecting objects in a closed container.

3. The final syllable contains a short vowel.

Not as bad as example 1, but for it to fully work, a syllable would have to be a set of separate phonemes (consonant and vowels) with well-defined boundaries (e.g., each separated by a glottal stop), which is never the case.

4. The egg lecithin used contains five different fatty acids.

Continuous, viscous fluid, a mix of chemical compounds, no well-defined boundaries, not the correct small spatial story for sets and elements.

5. The Tudor stable block contains an exhibition about the Battle of Quebec.

The stable block is probably more than a collection of exhibitions. We do have a container here, and the exhibition could be considered a subset contained within—the boundaries probably well defined. But the subset should consist of the same elements as the rest of the set (cf. the definition in Sect. 3.3.1), which is probably not the case.

6. The section on "Drama and Story" contains a number of brief practical examples.

This example would fit the set-theoretical small story but only if we assumed that it was divided into well-defined sections (there is not enough context information).

7. For all the joys that the world contains, reckon that love is the goldmine and those other things but gilded.

For this example to "work" the world would have to be a set of separate, well-defined elements like say "joys," "sorrows," etc. "Joys" could be considered a subset of which love would be an element. If only the boundaries of love and other feelings could be defined.

8. The positive charge in a nucleus is due to the several protons it contains.

If the nucleus were a set of noninteracting, separate elementary particles, the above example would fit the mathematical sense of set and element, and also the nuclear bomb would never be built.

9. Cornwall still contains a lot of tin.

If we imagined Cornwall as a set of atoms, all the atoms of tin, a distinctive chemical element, would be a well-defined subset (all the elements of S are elements of A, cf. Sect. 3.3.1). Fortunately, especially for those of us waiting for the next season of Poldark,[5] Cornwall is much more than that.

10. A basic slow-cooker is an earthenware pot, with a lid, set in a metal frame that contains a heating element.

A heating element is part of the metal frame, but to be a subset, it would have to contain the same elements, of which there is no

[5] http://www.imdb.com/title/tt3636060/, accessed 2017-10-10.

indication. Also, it should have well-defined borders, which may not be the case here.

All the "bad" examples (all of them except example 2) are also understood through image schemas, small spatial stories, and conceptual blending. This is how human mind works, "the way we think" to quote the title of the seminal book by Fauconnier and Turner (2002). But those other examples are understood through different image schemas and small spatial stories than the ones on which the standard set theory is built. What brought us to this conclusion? Mainly, the definition of a subset in Sect. 3.3.1. The only way of establishing whether A is a subset of B is making sure that A contains the elements that B also contains. If a subset was defined differently (or not defined at all, as in Leśniewski's mereology; see the next section), the other nine sentences could fit the theory. A large part of understanding mathematics is knowing which small spatial stories are "the correct ones." And this conclusion, as we will see in the next chapters, applies not only to the set theory but to all the other parts of algebra. Brilliant mathematicians, like the author of our handbook, Israel Nathan Herstein, always use "the correct" small spatial stories but mostly unconsciously,[6] so the students of mathematics are on their own. They must try to discover the right ones for themselves, often through the laborious process of trial and error.

More about the "correct" small spatial stories in the next section where we will also learn that mathematical sets are usually defined not through enumeration (they are too large for that) but by providing a condition, a shared property.

3.8 Defining Sets with a Condition and Russell's Paradox

> One final, purely notational remark: Given a set S we shall constantly use the notation $A = \{a \in S \mid P(a)\}$ to read 'A is the set of all elements in S for which the property P holds'. (Herstein 1975: 2)

[6] In fact, conceptual blending is always mostly unconscious—"These operations (conceptual blending, JW) -basic, mysterious, powerful, complex, and mostly unconscious-are at the heart of even the simplest possible meaning" (Fauconnier and Turner 2002: 6).

We learn that elements of sets can have properties and we can create sets of elements sharing the same property. Elements can be evaluated according to a property or a set of properties and a new collection can be created—for example, a set of all my pencils that are red. We have to say that Herstein's remark can hardly be classified as "purely notational," and the author of *Topics in Algebra* was of course aware of that but, most likely, did not want to confuse the reader at the very beginning of his handbook. Assuming that we can construct any set of elements sharing a given property has grave theoretical consequences which led to Russell's paradox (axioms of ZFC, the standard set theory, were created as a reaction to it). For example, let us imagine a set of all sets that have more than three elements. There is an infinity of such sets, like, say, a set of all cups, a set of all pencils, a set of all red hens, etc.—all containing more than three elements. Let us call it X. $X = \{A| \, o(A) > 3\}$, where o(A) (the order of set A) is the number of elements in A. Obviously X itself has more than three elements and therefore X is its own element: $X = \{set1, set2, set3, ..., X\}$. Try to imagine a collection that contains itself and some other elements beside it. Every set contains itself, $A \subset A$ and, in our experience with collections, nothing but itself. Sets like our X are called "improper sets." Let us now create a set of all "proper" sets, i.e., sets that are not like X, sets that are not their own element. There is an infinity of proper sets. The already mentioned set of all cups is a proper set, for example. How do we know it? Because this set is not a cup, and therefore not its own element. The question Bertrand Russell asked (1901) can be rephrased like this: is the set of all proper sets a proper set? Or is it its own element? And any of the two possible answers to this question lead to a contradiction. If we assumed that it was a proper set, it would have to be its own element (because it is defined as a set of all proper sets) and therefore improper. If we assumed it was improper, it would have to be its own element, and therefore it would have to be one of the proper sets (because it is defined as a set of all proper sets). It is surprising, isn't it, that such a basic, simple construction as the set theory leads to a contradiction. How is it possible? Especially considering the rigorous way the mathematical theories are constructed.

We will now try to answer this question and at the same time explain how Russell's paradox is connected to the main theme of this book (small spatial stories and conceptual blending in mathematics). In the first decades of the twentieth century, Russell's paradox was dealt with in two ways: (1) fixing the existing theory (Ernst, Zermelo) and (2) creating new set theories (Stanisław Leśniewski, Bertrand Russel). Zermelo created a set of axioms which arbitrarily forbid the construction of improper sets, while Leśniewski in his theory got rid of the element/subset duality we discussed in the previous section. In his theory element and subset are one and the same, and the whole theory, the mereology, and the theory of parts are built on only one image schema: the part/whole relation. What Leśniewski discovered is that Russell's paradox is the result of the clash between two image schemas (small spatial stories) of the container and part/whole, represented in elements and subsets. In the theory of parts, every set (class) is necessarily its own part/subset/element and therefore "improper" in Russell's terms. The set of all proper sets—the basis of Russell's paradox—is therefore nonexistent and considering its features makes no sense. Of course, Leśniewski could not have used the terms of modern cognitive science (image schemas, small spatial stories, conceptual blending); instead, he speaks of intuition and "turning to reality." This is how he comments on Zermelo's axiomatic solution:

> Architecturally sophisticated construction of Mr. Zermelo introduces to the set theory a number of unintuitive injunctions aiming at eliminating the paradoxes from mathematics. The question [...] of whether the set theory of Mr. Zermelo will ever lead to contradictions[7] is immaterial from the point of view of the intellectual torment caused by the reality-oriented imperative of the intuitive 'truth' of certain premises and accuracy of certain ways of reasoning [...]. From this point of view, the only method of solving the paradoxes is the intuitive eliminating of the errors of reasoning or premises leading to a contradiction. Unintuitive mathematics will not cure the deficiency of intuition. (Leśniewski 1913: 167, translated by JW)

Let us repeat that only one of the two types of containment in the set theory triggers Russell's paradox—the element, not the subset. It would be impossible to create the subset counterpart of a proper set. And this is because every set is by definition its own subset (sometimes referred to as the "improper" subset). When subset and element become one and the same, as in Leśniewski's mereology, all sets are their own elements (subsets) and therefore improper. When we eliminate

[7]Leśniewski anticipates here, for example, the famous Banach-Tarski paradox 11 years before it was discovered (in 1924).

one of the two clashing small spatial stories defining containment, the famous paradox vanishes.

If we look again at this section's introductory quotation, we will learn that "properties hold," which means that they are actors capable of exerting and maintaining force and of grasping. It is a small spatial story (of an actor "moving and shaking"). In the next section, we will analyze an example of set-theoretical proposition in search of more small spatial stories like the ones above.

3.9 Proposition, Proof, and Small Spatial Stories

For the definitions of the basic terms behind us, we are now ready for our first set-theoretical proposition and proof.

> PROPOSITION For any three sets, A, B, C, we have
> $A \cap (B \cup C) = (A \cap B) \cup (A \cap C)$.
> *Proof.* The proof will consist of showing, to begin with, the relation
> $(A \cap B) \cup (A \cap C) \subset A \cap (B \cup C)$ and then the converse relation
> $A \cap (B \cup C) \subset (A \cap B) \cup (A \cap C)$.
> We first dispose of $(A \cap B) \cup (A \cap C) \subset A \cap (B \cup C)$. Because
> $B \subset B \cup C$, it is immediate that $A \cap B \subset A \cap (B \cup C)$. In a similar
> manner, $A \cap C \subset A \cap (B \cup C)$. Therefore
> $(A \cap B) \cup (A \cap C) \subset (A \cap (B \cup C)) \cup (A \cap (B \cup C)) = A \cap (B \cup C)$.
> Now for the other direction. [...] (Herstein 1975: 4)

In fact, the above is just the first half of the proof but enough for us to take a closer look at the linguistic and conceptual patterns. The proposition reminds us of the familiar feature we remember from arithmetic, for example, $2*(3 + 4) = (2*3) + (2*4)$, the so-called distributive property, more specifically, the multiplication being distributive over addition or, in more familiar terms, expanding brackets. And this is exactly what it is, with addition and multiplication of numbers replaced with union and intersection of sets.

The first line of the proposition contains "for any three sets, A, B, C, we have," which is of course synonymous with "for any three sets, A, B, C, the following is true." When a statement is true, "we have it." We are in possession and we are the proud owners of all true statements. And true statements are things we have. We probably keep those things in a safe somewhere or in a gallery and look at them proudly, thinking they are ours, we are the owners. And sometimes other people give us new items for our collection, they share them with us, we reciprocate, etc. What we have here (!) is of course another small spatial story projected into the narrative of mathematical proof. And our small spatial story is then enriched with new elements, for example, "we first dispose of...," which in this context is synonymous with "we prove that." The statement to prove is an object we dispose of, we get it out of the way, so that we can get access to other objects on our path of mathematical proof and dispose of them too until we reach the spot where it says "QED—your work is done." Our small spatial story expands, and if it does not stop, soon we will have to call it a medium-sized spatial story. In the little fragments of text inserted between the mathematical symbols, we also find "it is immediate that," which we are to interpret as "one follows from the other and it is easy to understand." But the ease, the low level of effort, is expressed metonymically through the (short) time it takes us to understand that the next statement follows from the previous one. And as we already observed, in our small spatial story of mathematical proof, we move along the path, and sometimes we have to move back: "and now for the other direction." And on our way back, we will be disposing of other objects, storing the truths in our collection of true statements, and clearing a wider path to the mathematical truth. Finally, the small spatial story will have run its course. "The end" (or QED) will be displayed on the screen and the credits will roll.

We have just demonstrated that mathematical proof is understood in terms of small spatial stories of possession (ownership), collecting, removing obstacles, and moving along a path.

Summary

In this chapter we have discovered that the set theory, just like One Thousand and One Tales of Scheherazade, contains objects, actors, actions, image schemas, and conceptual blending.[8] Examples are listed in Table 3.2 below.

The above "mental patterns" are structured as follows:

1. A small spatial story (actors moving in space and manipulating objects) is part of the mathematical narrative.

[8] Mark Turner (1996: 9) calls the above "mental patterns of parable."

2. The small spatial story becomes one of the inputs of a conceptual integration network (in the mind of the reader).
3. The conceptual integration network is where mathematical meaning is constructed.

When we say that "the narrative contains blending and small spatial stories," it is a metonymic shortcut of course—they are mental patterns, which can be triggered by the narrative but certainly do not reside there. Unlike the small spatial stories and their building blocks, the image schemas, which seem to be everywhere, blending is not easy to spot, being a mostly unconscious process. But we found traces of it in the mathematical narrative with perhaps the most tell-tale example being the equality of sets (Sect. 3.3.2). We need two identical tokens of a unique set to compare and to conclude they are indeed the same. The set has to be "one and many" at the same time, which is indicative of the bi-directional projection between the blend and the input spaces of the conceptual integration network.

In the next chapter, we will examine the second pillar on which the edifice of algebra rests—the mapping.

Chapter 4
Mappings

4.1 Overview

As in the previous chapter, we will continue to look for small spatial stories and conceptual blending to see how those mental patterns contribute to the effectiveness and fecundity of mathematics. The subject of focus of this chapter is mathematical mapping, described in the quotation below as "the single, most important and universal notion." We will find that at least three small spatial stories are prompted in our handbook as a way of constructing the meaning of mapping. Or, to use the terms introduced in Sect. 2.2, three small spatial stories become input spaces in the conceptual integration network where the meaning of mathematical mapping emerges. As we mentioned above, the input small spatial stories are always easier to spot than the mostly unconscious process of conceptual blending, but we will see several traces of it as well. A rather surprising secondary conclusion of the present chapter is that the official, "rigorous" definition of mapping as a set of ordered pairs is in fact fully circular, which—taking into account the crucial importance of mapping—should result in the whole discipline being completely barren. Instead, mathematics thrives, and its effectiveness is constantly confirmed by new applications in all areas of science and industry. In the following section will be able to see how conceptual blending and its input small spatial stories contribute to these accomplishments.

J. Woźny, *How We Understand Mathematics*, Mathematics in Mind,
https://doi.org/10.1007/978-3-319-77688-0_4

4.2 The Mapping as "a Carrier"

> We are about to introduce the concept of a mapping of one set into another. Without exaggeration this is probably the single most important and universal notion that runs through all of mathematics. It is hardly a new thing to any of us, for we have been considering mappings from the very earliest days of our mathematical training. When we were asked to plot the relation y = x^2, we were simply being asked to study the particular mapping which takes every real number onto its square. (Herstein 1975: 10)

One set is mapped "into" another—the preposition clearly suggests motion that ends in a container. And $y = x^2$ function (mapping) "takes every real number onto its square." The mapping takes objects "onto" other objects. The scenario is all too clear: the mapping sees a number, takes it, moves with it along a path, and then drops it on a specific spot. The mapping then retraces its steps and goes looking for other numbers to carry and continue until the job is done and all the numbers are carried onto prescribed spots. And those spots are other numbers. So they are already taken. The mapping (actor/agent) has then no choice but to place the carried numbers "onto" those other numbers that are already there. So those target spots have now two numbers on them piled one on top of another. We will return to the small spatial story of the carrier, but, in the next section, we will first focus on the official definition of a mapping.

4.3 The "Rigorous" Definition, Ordered Pairs

Let us take a look at the "rigorous" (as Herstein describes it) definition of a mapping: "If S and T are nonempty sets, then a mapping from S to T is a subset, M, of S × T such that for every s ∈ S there is a unique t ∈ T such that the ordered pair (s, t) is in M" (10). Where is all the motion now? The mapping is still "from S to T," but otherwise we learn that it is a subset, and subsets don't move, do they? And this subset contains "ordered pairs." So the concept of order is added and never defined (just like sets and elements before). It is not that Herstein forgot to define it—it is just another of those "primitives" or primary concepts that we are supposed to know from our everyday experience.

So what is "order" exactly? Let us think of some examples. We say "ordered from left to right." And "left and right" are directions relative to our current location and of course orientation of our body. They (left and right directions) depend on our bodies being asymmetrical, having a well-defined front and back. We also say "on my left," "on your right," etc. So far we have established that order might have something to do with location of objects relative to our bodies. Is there a definition of left and right somewhere? Unfortunately not, at least not in mathematics. Left and right depend on a convention, just like language (to a degree)—they are not universal. There must be a community of people first, who agree that this direction (relative to my body orientation) will be henceforth called "left" and the other one "right." And then, when the convention is established, we can order things from left to right.

Is order only a spatial concept then? Certainly not. We also say "I saw a sheep and then a cow." So it seems we can have "temporal order" as well. On the other hand, the concept of time is projected from space. How else can we order things then? We can say, for example, "these are my red pencils and these are my blue pencils and I keep them in separate boxes." So, not space but, more generally, objects differing from one another, objects having different features (e.g., of location, shape, time, or color). Coming back to our "ordered pair" (s,t), s (the mark on the screen or on paper) is more to my left, and t (also the mark on the screen or on paper) is more to my right. The s and t letters have different locations (just like pencils can have different colors), and this is the feature we can order them by.

A crucial mathematical concept depending for its sense on the position of two squiggles on paper. Is it possible? Can't we find something more profound, more mathematical? Unfortunately not. As we said, order (like set and element) is a "primitive" (undefined) concept. One of many such concepts that mathematicians leave for us to figure out for ourselves. And we tried, coming to the conclusion that the ordered pair of (s,t) is a mapping of two different spatial locations (of two squiggles on paper or on the screen) on the elements of the set {s, t}. So in fact, as we see, to understand the mathematical concept of mapping, we have to use an undefined concept that relies on mapping. Conundrum. Circularity. Concepts defined with oblique reference to themselves. In the next section we will explain how small spatial stories and conceptual integration create a way out of this problem.

4.4 Circularity of the "Rigorous" Definition and Conceptual Integration

The question we should ask now, is it just sloppy work by mathematicians. Is there a better way of defining mapping? Can we have better definitions? Perhaps I should write a new algebra handbook that would be so much better than all those old books like Herstein's. And the answer is of course no. If we believe that mathematics is a product of the human mind (and we do, and we have been finding proof of it again and again in this chapter), then it must be "flawed" in the same way that human mind is "flawed." So what if the definition of mapping is circular? Circularity is one of the most obvious features of the human thought process. Canonically, two inputs are mapped into the blend where a new meaning/content/structure emerges which is then mapped BACK to input spaces, and no one said this cycle is repeated only once. Is this not circularity? Where the input concept/meaning is modified by itself via the blend? Of course it is. And it is not a "flaw" at all—it is simply the way this wonderful conceptualizer of ours works. And we have just found a trace of it in the "rigorous" definition of a mathematical mapping.

Now that we have mapping "defined" (the circular definition based on the mapping of the ordered pair), we could define ordered pair "rigorously" as, for example, the following mapping form {1,2} into {s,t}. (s,t) = {(1,s), (2,t)}. Of course this "definition" is explicitly circular—an ordered pair defined as a set (mapping) of ordered pairs, but, as we mentioned above, that is how our mind and mathematics as its product (often) works. We could of course use the reverse mapping from {s,t} into {1,2}. (s,t) = {(s,1), (t,2)}, and it would work as well (as long as it is a one-to-one mapping). And this is what numbering objects in a set is—a mapping from this set into a set of numbers (indexes) or vice versa. And when we have this mapping, we can say s is number 1 and t is number 2. Instead of numbers, we could use any other set of indexes, like {left, right}, {red, blue}, {sooner, later}, etc.

Let us remember our first glimpse of a mathematical mapping, a few paragraphs above: "the particular mapping which takes every real number onto its square." Notice that there is no circularity there, but instead we have an actor carrying numbers to appointed (mapped) places. And by doing this, this actor was also ordering. This exactly is what ordering is—carrying objects from one place to another (putting empty cups from my desk into the dishwasher, pens into the drawer, bits of paper into the trash can, etc.). Mathematicians used the concept of "order" to define mapping although conceptually mapping and order are one and the same. Of course, if I carried bits of paper from my desk into the dishwasher and put empty cups into the drawer and my perfectly good pens into the trash can (happens to all of us sometimes, doesn't it?), I would not be strictly speaking "ordering," so perhaps it would be better to conclude simply that ordering is a type of mapping, and mapping is carrying objects from one place to another. And again there is no circularity. We have achieved a certain order of thought and peace of mind. Our concepts are ordered and safe now. *Everything is where it should be, and everything is in its place* (this would make an excellent definition of "order").

Let us go back to the language of "pure mathematics" and repeat the "rigorous" (circular) definition of mapping: "If S and T are non-empty sets, then a mapping from S to T is a subset, M, of S × T such that for every s ∈ S there is a unique t ∈ T such that the ordered pair (s, t) is in M." In Herstein's words: "probably the single most important and universal notion that runs through all of mathematics" (1975: 10). We have demonstrated above that this crucial "definition" is circular, but we have also mentioned that circularity is sometimes unavoidable, a consequence of the way human mind works, a consequence of conceptual integration, the most basic, constant, usually unnoticed, and yet most profound of our mental abilities: "The human spark comes from our advanced ability to blend ideas to make new ideas. Blending is the origin of ideas [...] Blending, I claim, is the big lever of the cognitively modern human mind" (Turner 2014: 9). And, as we observed above, circularity is built into blending because the mapping between the inputs and the blend is bi-directional, which means that the input concept is "modified by itself" via the blend.

The ordered pair definition of the mapping is presented as the official, rigorous one, but as we will learn in the next chapter, it is for some reason "almost never used," and another "way of thinking about it" is preferred.

4.5 He Small Spatial Story of the Matchmaker

We have hypothesized above how the process of conceptual blending can help mathematicians live with, and even go on and do amazing things with, mathematics that is based upon a definition of mapping which is evidently circular. So let us enjoy the circular definition again, this time with Herstein's comment on it:

> DEFINITION: If S and T are nonempty sets, then a mapping from S to T is a subset, M, of SxT such that for every $s \in S$ there is a unique $t \in T$ such that the ordered pair (s, t) is in M.
>
> This definition serves to make the concept of a mapping precise for us **but we shall almost never use it in this form. Instead we do prefer to think of a mapping as a rule which associates** [emphasis added] with any element s in S some element t in T, the rule being, associate (or map) $s \in S$ with $t \in T$ if and only if $(s, t) \in M$. We shall say that t is the image of s under the mapping. (Herstein 1975: 10)

We have already established beyond doubt that this definition does not "make the concept of mapping precise"—on the contrary, this definition of mapping is based on the (never-defined) concept of ordering, and ordering is mapping. Full circle. Let us be honest here—it is a terrible definition. But Herstein, confirming (or at least not disproving) our speculations in the previous section, uses the phrase "we do prefer to think of a mapping as [...]." So, yes, we have this definition, but there are many ways of "thinking about it." What Herstein is telling us may be interpreted as "blend at will—there are many ways to construct the meaning of mathematical definitions." And then he uses the phrase "rule which associates." And the circularity magically disappears! But we are back to the "actor scenario"—a small spatial story. But this time we have a different actor. Instead of carrying objects from one place to another, this actor (and her name is Rule) only "associates" (connects) objects. No more heavy lifting— this is a white-collar job, a matchmaking agency: "Hello, s? We have found an excellent partner for you. His name is t." And thus a pair of (s,t) is created and entered into Ms Rule's notebook before she proceeds to phone the next client, etc.

As we will find in the next section, "the matchmaker" and "the carrier" are not the only small spatial stories contributing to the meaning of mathematical mapping.

4.6 Definition by Graph, the Small Spatial Story of a Hiker

Defining a mapping as a set of ordered pairs is sometimes referred to as "definition by graph." This is how Herstein introduces it:

> Let us motivate a little the definition that we will make. The point of view we take is to consider the mapping to be defined by its "graph." We illustrate this with the familiar example $y = x^2$ defined on the real numbers S and taking its values also in S. For this set S, S x S, the set of all pairs (a, b) can be viewed as the plane, the pair (a, b) corresponding to the point whose coordinates are a and b, respectively.[1] In this plane we single out all those points whose coordinates are of the form (x, x^2) and call this set of points the graph of $y = x^2$. (Herstein 1975: 10)

So, we do not have to think of a mapping a set of ordered pairs (which leads to circularity). Instead we can think of it as a geometric shape, "the graph", a set of points on a plane which are "singled out." But this definition of mapping —the shape, the graph—is not static at all because it prompts for action. In the primary school we are taught how a line graph works: "if you want to know what the value of the function (mapping) for x is, draw a line perpendicular to the x-axis until it crosses the graph and then from that point, another line, parallel to the x-axis, until it crosses the y-axis—and this is where your y is."

We can imagine point *x* travel across the plane, and the points of the graph work as a map that tells us where to "turn" to find y (Fig. 4.1). This is another spatial/dynamic story we choose arbitrarily to refer to

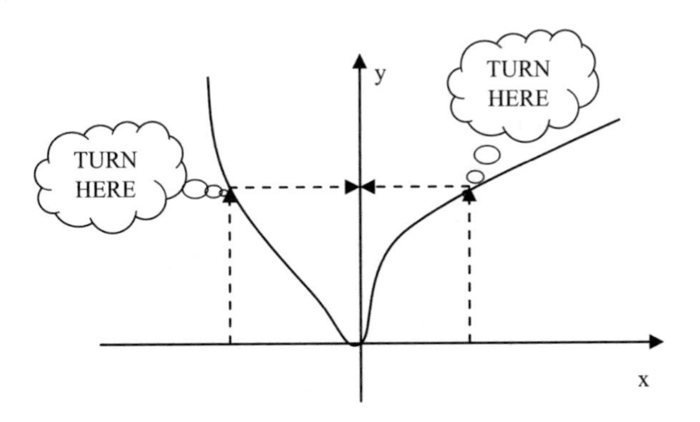

Fig. 4.1 Graph of a function—the dynamic scenario of *the hiker*

[1] Herstein makes it sound rather easy, but "Western" geometry did not make this association until the seventeenth century (Descartes, Fermat). The mapping of SxS onto the plane is the base of analytic (or "Cartesian") geometry.

as "a hiker" (the story of a person who needs a map to know where to turn to reach her destination).

In the next section we will consider the question how exactly the small spatial stories help us avoid the circularity of the "rigorous" definition.

4.7 Structured Small Spatial Stories vs. Circularity of the Definition

The small spatial stories so far (the carrier, the matchmaker, the hiker) seem to work very well as the conceptual ground for mathematical mapping, and they do not result in circularity. And this is because they are structured and inherently ordered, and there is no need to introduce "order" or "ordered pair" as a primitive, undefined notion. Every small spatial story contains actors/agents, actions, and objects/patients. Actors/agents carry out the actions according to a certain scenario. The carrier transports x to y, the rule associates x with y, and the hiker travels from x to y. Mathematical mapping, for its integrity, must forever remain a projection of (a conceptual integration network built on) a small spatial story. None of those stories, let us repeat, require pre-ordered pairs (the source of circularity) because the scenario introduces order. For example, when the carrier transports an object (number) to a certain spot (also a number), we know which of these numbers is the x (argument) and which is the y (value) because they occupy different slots in the scenario—x is the object being carried, and y is the goal of the path. Similarly, in the matchmaker story (Ms Rule), x is the client and y is the partner that the matchmaker finds for the client. The Cartesian plane story is even more explicitly structured—x is on the horizontal axis, and y is the point we can find on the vertical axis if we turn at the right point. Mathematical mapping can only be understood through a dynamic scenario or as a blend of a number of dynamic scenarios—in other words, through small spatial stories. I do not think it is a coincidence that Herstein declares the static (circular) definition to be "almost never used" and prompts for a dynamic scenario instead. This is what quoted above Stanisław Leśniewski, the creator of the "theory of parts" (mereology), meant as "the reality-oriented imperative of the intuitive truth" (1913: 167).

Summary

Table 4.1 summarizes the traces of small spatial stories (actors, objects, actions) and conceptual integration we discovered by analyzing the description of mapping in Herstein's *Topics in Algebra* (1975). We have found three distinctive small spatial stories contributing to the construction of the meaning of mapping, and we have learned that mathematicians can switch between the small spatial stories they use or even have their preferred ones (cf. Sect. 4.5). We have also come to a rather surprising conclusion that the primitive concept of "order" ("ordered pair") is in fact a mapping, which renders the official definition of a mapping circular; however, we argued that circularity is an inherent feature of human thought process, due to the bi-directionality of mapping in the conceptual integration network. An ordered pair is only one of the possible inputs in this network, the others are the set, and the small spatial stories mentioned above. The seemingly boundless fecundity of mathematics (and other areas of human creativity) confirms that circularity does not equal redundancy.

Now that we have discussed both the set and the mapping; we are ready to move on to the theory of groups because the shortest characterization of a group is "a set with a mapping." In fact, historically, the prototype of a group was a set of mappings ("permutations"), the so-called "symmetry group."[2]

Table 4.1 Elements of small spatial stories and traces of conceptual integration found in the narrative of the mathematical mapping

Objects	Numbers, elements of sets
Actors	The carrier, the matchmaker, the hiker
Actions	Carrying, associating, moving from x to y
Image schemas	Source-path-goal, compulsion, link, matching, superimposition, diversion, object, container, process
Conceptual blending	Input spaces of the conceptual integration network can contain ordered pairs, points on the plane, carrying objects, associating objects, motion along a path. The circularity of the "rigorous" definition may reflect the circularity inherent in the process of blending. In any conceptual integration network, the mapping is bi-directional

[2] Symmetry group is the set of all one-to-one mappings (also called "permutations") of a set S onto itself, in Herstein's handbook marked as A(S).

Chapter 5
Groups

5.1 Overview

As in the previous chapters, we will follow the narrative of algebra to see how mathematical meaning emerges from small spatial stories and conceptual blending. This time we will focus on the narrative of the group theory considered to be one of the most beautiful areas of algebra (especially for the finite groups). This is how the chapter on group theory begins in Herstein's handbook:

> In this chapter we shall embark on the study of the algebraic object known as a group which serves as one of the fundamental building blocks for the subject today called abstract algebra. [...] In abstract algebra we have certain basic systems which, in the history and development of mathematics, have achieved positions of paramount importance. These are usually sets on whose elements we can operate algebraically-by this we mean that we can combine two elements of the set, perhaps in several ways, to obtain a third element of the set. (1975: 26).

We learn that an algebraic operation (e.g., like addition) can "combine two elements to obtain a third element." We had "combining" in Chap. 3 but as an operation on sets (the union of sets), not elements. The elements, objects in a collection, could never be combined. The notion of "a union of elements" simply does not exist in the set theory. This time it is a different type of combining—combining through mapping by creating the so-called ordered pairs we discussed in Chap. 4. Another actor, a craftsman again, is able to combine elements. We will learn later that one of the basic features of algebraic

J. Woźny, *How We Understand Mathematics*, Mathematics in Mind,
https://doi.org/10.1007/978-3-319-77688-0_5

operations (like the already mentioned addition, or multiplication, division, etc.) is that they "stay within the set." For example, if we add two integers, the result is also an integer. Interestingly, when combining two numbers (2 + 3 = 5), we "obtain" a third number. So we "obtain" an already existing element, or rather we obtain a token of it, which then becomes one with itself. And we have seen this tokenization in the previous chapter, when we discussed conceptual blending in connection with the "=" sign—the relation of equality (identity). And in the same sentence, we learn that there are "different ways" of combining. And each of those ways will require a different specialized craftsman. We can call them the adder, the subtractor, the multiplier, and the divider. New actors and traces of conceptual integration in one sentence—a promising start of a new chapter.

Let us enjoy one more fragment of Herstein's introduction to the group theory, in which he explains what "good mathematics" is.

> We should like to stress that these algebraic systems and the axioms which define them must have a certain naturality about them. They must come from the experience of looking at many examples; they should be rich in meaningful results. One does not just sit down, list a few axioms, and then proceed to study the system so described. This, admittedly, is done by some, but most mathematicians would dismiss these attempts as poor mathematics. The systems chosen for study are chosen because particular cases of these structures have appeared time and time again, because someone finally noted that these special cases were indeed special instances of a general phenomenon, because one notices analogies between two highly disparate mathematical objects and so is led to a search for the root of these analogies. (Herstein 1975: 26).

Axioms must have "naturality." Have we finally found an algebra handbook based on the concept of embodied cognition, the central idea of cognitive linguistics, and cognitive science?[1] The next sentence certainly does not quash this hope: "experience of looking at many examples," "meaningful results," etc. But, unfortunately, when we read the reminder of the above passage, there is no hope left for Herstein's book joining the canon of cognitive science. The passage ends with "search for the root of these analogies." And by "root," Herstein means axioms. He believes mathematical truths are already out there (like Platonic forms); we just have to find the preexisting rule—the correct axiom.

---

[1] cf. Sect. 2.2.1

Mathematical beauty is a derivative of understanding and of grasping the meaning of axioms, theorems, and proof. Mathematical beauty follows mathematical truth. And our book is about the construction of mathematical meaning (based on conceptual blending, with small spatial stories as the input), so there is a clear connection. And in this chapter, we will try to grasp them both: truth and beauty of mathematical groups. The two previous chapters prepared us for it because, as we will see in the following section, the definition of a group is based on the notions of a set and a mapping.

5.2 The Definition of a Group and the Story of the Matchmakers

This is how a group is defined:

> DEFINITION A nonempty set of elements G is said to form a group if in G there is defined a binary operation, called the product and denoted by ·, such that
>
> 1. $a, b \in G$ implies that $a \cdot b \in G$ (closed).
> 2. $a, b, c \in G$ implies that $a \cdot (b \cdot c) = (a \cdot b) \cdot c$ (associative law).
> 3. There exists an element $e \in G$ such that $a \cdot e = e \cdot a = a$ for all $a \in G$ (the existence of an identity element in G).
> 4. For every $a \in G$ there exists an element $a^{-1} \in G$ such that $a \cdot a^{-1} = a^{-1} \cdot a = e$ (the existence of inverses in G). (Herstein 1975 : 28)

We learn that "a set forms a group." So in this case, the set is an actor or perhaps rather a metonymic reference to an actor who forms something out of a set. For this actor, the set is a rough material out of which something new is formed. The craftsman (agent) has to perform certain actions before the rough material becomes a finished product. And the group-forming procedure includes defining a binary operation, which has the specified features. The binary operation is a mapping, which we studied at length in the previous chapter. A mapping, as we established and Herstein (1975: 10) confirmed, can be "thought of" in many different ways. Herstein's preference was to think of it as a "rule which associates elements." This associating rule is another actor, which we chose to call "the matchmaker." In our mathematical narrative, we now have an actor/agent—the matchmaker. But not just any kind, the interview procedure is very strict.

This matchmaker must be able to do her job (connecting elements) according to four given rules—the group axioms.

Before we concentrate on the rules to follow, let us just add that, in fact, a second actor is also involved in the process of group forming—another matchmaker. To avoid confusion and also tell our tale of group formation in chronological order, we will call her "matchmaker1." The job of matchmaker1 is to match all the elements of the rough material set G (not yet a finished group) in ordered pairs to form the Cartesian square G × G. And when this semifinished product is ready, matchmaker2 will associate each ordered pair with another citizen (element) of the set. To summarize, two actors/agents operate on the rough material of the set G: matchmaker1 and matchmaker2. Each of them has specialized tasks. Matchmaker1 associates all the elements of G into ordered pairs, and matchmaker2 associates each of those pairs with another element. Of those two, matchmaker2 has the toughest job—not only to perform the matching but to do that according to four strict rules.

The first rule matchmaker2 must follow is never to look for a match outside the set. The set is "closed" with regard to her actions; G is an isolated community allowing only "internal" marriages (or rather triangles)—exogamy strictly forbidden.

Rule number 2, "the associative law" of the realm G is, as laws often are, quite complicated although the notation is very simple and familiar; we associate it immediately with multiplication. And we remember from primary school that we can use the brackets however we want without changing the "result." What does it mean for matchmaker2? The match for (a,b) is (a·b), and now this match can also be matched with another element—c. And the result of this new match is (a·b)·c. But the order of matching can change, and the couple (b, c) can be matched first with (b·c), and then the result of matching element a with it would be a·(b·c). And it has to be the same as the result of the previous match, where the first two clients (a and b) were matched first. So, whenever there is a group of three clients—a,b,c—matchmaker2 must remember that the order of matches does not matter; the final match—a·b·c—must always be the same.

Rule number 3 requires that there exist a special element in G (an equivalent of 1 for multiplication or 0 for addition) such that the couple (a,e) or (e,a) is matched to a again. Element e is "Mr. Cellophane" (from *Chicago*[2]); you can "look right through him, walk right by him, and never know he is there."

Our Mr. Cellophane also features prominently in Rule 4. Matchmaker2 must always be able to find a mirror-opposite of element a—an anti-a—such that when they are matched as a pair, the result is null, void, and cellophane (0 in addition, 1 in multiplication, etc.).

The next chapter contains the first hint of mathematical beauty for finite groups.

5.3 Abelian Groups, Finite Groups, and the Beauty of Mathematics (Part 1)

Definitions of abelian groups and finite groups follow, with Herstein's commentary.

> DEFINITION A group G is said to be abelian (or commutative) if for every a, b ∈ G, a·b = b·a. A group which is not abelian is called, naturally enough, non-abelian; having seen a family of examples of such groups we know that non-abelian groups do indeed exist. Another natural characteristic of a group G is the number of elements it contains. We call this the order of G and denote it by o(G). This number is, of course, *most interesting when it is finite* [emphasis added]. In that case we say that G is a finite group. (Herstein 1975: 28).

On the same page, we will find that "it is not surprising that for every nonempty set S the set A(S) is a group. Thus we already have presented to us an infinite source of *interesting*, concrete groups" (ibid.). A(S) is a set of all possible one-to-one mapping of S onto itself. And on the previous page, we read: "Amongst mathematicians neither *the beauty* nor the significance of [...] groups is disputed" (ibid.: 27). I have used emphasis in the above quotations to draw the reader's attention to the concept of mathematical beauty—in our case, the beauty of groups, especially finite groups. And Herstein is telling us that it is not disputed among mathematicians. Only they can see it; he seems to be claiming. But the fact that mathematics may be considered beautiful—and in a moment we, mere mortals, will try to have a glimpse of it as well—does not contradict the main claim of this book (and many other sources mentioned so far) that mathematics is

[2] http://www.imdb.com/title/tt0299658/, accessed 2016–12-07

a product of the human mind, because various products of the human mind, like art, literature, or music, are considered beautiful as well. And we do not need a separate proof that mathematicians are homo sapiens (cf. Lakoff and Nunez 2000: 1), so perhaps if they can see the beauty of mathematics, so can we? After all, one does not have to be an artist to see the beauty in Fig. 5.1 below.

Where does the beauty of finite groups dwell? And, more importantly, how is it connected to our main subject? In this book, we are trying not to skip ahead; notice that (almost) all quotations from Herstein's handbook appear in chronological (increasing page number) order. And to answer these questions, we need the concepts of a subgroup, a coset, normal subgroup homomorphism, and most of all, Lagrange's theorem. Therefore, the subject of mathematical beauty will be continued in Sects. 5.7, 5.8, and 5.9.

The next section contains Herstein's remarks on the objective nature of mathematics and also an apology for his (excellent in our opinion) book being in parts "rather dull." And of course, we humbly join him in this apology now but only with reference to the present volume.

Fig. 5.1 One of the Upper Paleolithic Chauvet cave paintings (https://www.theguardian.com/artanddesign/jonathanjonesblog/2014/mar/21/the-10-greatest-works-art-ever, accessed 2016-12-14)

5.4 On the Objective Nature of Mathematics

We (there are evidently some identity, or uniqueness, issues here as the book has one, singular author—very relevant for the next section) are austerely bent on studying the language of "pure mathematics". But now and then—please forgive us, dear reader—we cannot help ourselves and quote one of the "purely narrative" (the quotation marks equally justified when we use them for "pure mathematics"—of course we have narrative in both cases) fragments of Herstein's excellent algebra handbook that we have come to like very much. So here it comes:

> We have now been exposed to the theory of groups for several pages and as yet not a single, solitary fact has been proved about groups. It is high time to remedy this situation. Although the first few results we demonstrate are, admittedly, not very exciting (in fact, they are rather dull) they will be extremely useful. Learning the alphabet was probably not the most interesting part of our childhood education, yet, once this hurdle was cleared, fascinating vistas were opened before us. (Herstein 1975: 33)

Apparently, the mission is to "find facts about groups." Never mind that they have just been defined and called into being by virtue of manmade definition.[3] We are on a safari now, surrounded by exciting animals that we want to observe closely and "find facts about." Yes, mathematics is out there—Herstein is saying—finished, perfect, not at all a product of the human mind. Groups would be there whether we defined them or not. Numbers and matrices, mappings, sets and elements, points on a plane, the parabola, all the equations, 1.7, 1.71, and even 8.08, they would all be there even if we weren't. And how kind it is of an author of an advanced algebra handbook to worry about it being sometimes "not very exciting" or even "rather dull"? But, on the contrary, the sole (one) author of this book—who suffers from identity problem and now is looking at himself from above, referring to himself in the third person—disagrees. The next bit of "pure mathematics" is truly mind-boggling.

---

[3] It is true that groups existed before the "modern" (XIX c.) definition was created by Cayley (as sets of permutations, A(S)—the symmetry groups), but those were also defined and could also be described as "manmade."

5.5 The Uniqueness of the Group Elements and Conceptual Blending

LEMMA If G is a group, then
a. The identity element of G is unique.
b. Every $a \in G$ has a unique inverse in G. (Herstein 1975: 33)

The identity element in a group (1 for multiplication, 0 for addition, etc.) is unique. So there is only one "1" and only one "0." There can't be two (or more) 1's or two (or more) 0's. And the same applies (from point b) to 5 and −5, for example. −5 is unique. But, since 5 is the inverse of −5 (with respect to addition), 5 is also unique. And the same applies to all integers (they are a group) of course. How can we explain this equation then: $1 + 1 = 2$. Are there two ones on the left? Or are we hallucinating? Or take his one: $1 \cdot 1 = 1$. Three ones? Or just one? Let us remember the definition of a group: $1 + 1$ is an example of a result of an action of a binary operation (matchmaker2) who associated the "ordered" (yes, exactly!) pair of (1,1) with number 2. So the "uniqueness" problems started then, at the level of definition, in the process of the "forming" of the group. Who is responsible for this mess? Well of course—there is no doubt—it is matchmaker1. It was her job to create the Cartesian square of $G \times G$—the set of all ordered pairs (a,b) where both a and b are elements of G. And one of those pairs is (1,1). But look at the symbol for the Cartesian square—$G \times G$. How many sets G are there? Apparently, at least two. But they are the same, unique set. Two of the same. How could we have missed it? Easily. We already discussed it in Chap. 3 when we focused on the equality sign "=." Two input spaces containing tokens connected with identity relation, tokens that in the blend space are compressed into a unique entity. And therefore an object is "one and many" at the same time because of course the input spaces do not "disappear" or become "disconnected" but instead become part of conceptual integration network which is then "run from the blend." To quote one of the conceptual blending theory governing principles (the web principle): "Manipulating the blend as a unit must maintain the web of appropriate connections to the input spaces" (Fauconnier and Turner 2002: 331; cf. also Table 2.2 in Sect. 2.2.5). Given a single and unique set G, say {Adam, Jacek}, we construct (matchmaker1 does in the process of forming the group) another token of it and then build G × G = {(Adam, Adam), (Jacek, Jacek), (Adam, Jacek), (Jacek, Adam)}. The Cartesian square set G × G contains pairs of "unique elements"—the (a,a) pairs—the so-called diagonal subset. In our case those are (Jacek, Jacek) and (Adam, Adam). We construct ordered pairs of one unique entity. There is nothing unusual about it; we do it all the time, mostly unconsciously.[4] And we have just found another trace of it in the narrative of "pure mathematics." And I have just created an ordered pair of myself, to facilitate the writing process and move on at a faster pace to the next bit of mathematical narration.

> Before we proceed with the proof itself it might be advisable to see what it is that we are going to prove. In part (a) we want to show that if two elements e and f in G enjoy the property that for every a E G, a = a · e = e · a = a · f = f · a, then e = f. (Herstein 1975: 33)

It is interesting to notice how the problem of "one and many at the same time" is handled in the above quotation. It will be an indirect proof, in which the assumption that the opposite of the proposition is true leads to a contradiction. What is assumed to be two different elements—e and f—will be proven to be one element e = f. The contradiction then lies in the contrast of two different elements vs. one element. In other words, what Herstein is obliquely telling us is when e = f they are ONE element. And that there can't be TWO elements that are equal. Otherwise, there would be no contradiction and point a of our lemma would not be proven.

And yet group G, as we have seen above, is built on a Cartesian product of G × G, which contains, for example, the (e,e) pair, an ordered pair of one element (there can't be two, and it can be proved by contradiction). So in this case (and for all other (a,a) pairs of the G × G diagonal subset), an ordered pair contains one unique element, not two. But somehow, this is not a contradiction.

In the following section, we will take a closer look at the proof of the second part of the above lemma—the uniqueness of the inverse element.

---

[4] George Mikes did it consciously, however, when he famously wrote "An Englishman, even if he is alone, forms an orderly queue of one."

5.6 The Force Dynamics of Mathematical Proof

The proof of part b of the "uniqueness" lemma follows:

> Rather than proving part (b), we shall prove something stronger which immediately will imply part (b) as a consequence. Suppose that for a in G, a· x = e and a ·y = e; then, obviously, a· x = a ·y. Let us make this our starting point, that is, assume that a· x = a ·y for a, x,y in G. There is an element b E G such that b ·a = e (as far as we know yet there may be several such b's). Thus b · (a· x) = b ·(a ·y); using the associative law this leads to x = e · x = (b ·a) · x = b ·(a· x) = b · (a ·y) = (b ·a) ·y = e ·y = y. We have, in fact, proved that a· x = a ·y in a group G forces x = y. (Herstein 1975: 34)

Instead of a proof of part b, we will have something "stronger." In Chap. 3 we already described the small spatial story that is projected onto the process of mathematical proof. The actor—let's give her a name, *the truth collector*—went on her way removing obstacles and storing them at the same time in her repository of mathematical truths on her way to a QED goal. And now we learn that a proof possesses strength and is therefore capable of forceful interaction. Weaker (lesser) proofs "follow" from "the stronger ones." A proof, which is at the same time an element of the collection owned by the truth collector, can "force" another truth/proof. We have forceful interactions between mathematical proofs, and using one of them leads to consequences—the balls hit other balls which then start to move and on their way force other balls into motion until finally the red ball (the proposition) goes straight into the pocket. Yes, it is a billiard ball game of force and motion, of objects interacting with other objects, exerting force, and causing motion, yet another small spatial story and its projection in the narrative of mathematics.

In the next section, we will continue our pursuit of mathematical beauty, after introducing the concept of a subgroup and formulating the famous Lagrange's theorem.

5.7 The Subgroups, Lagrange's Theorem, and the Beauty of Mathematics (Part 2)

> In general we shall not be interested in arbitrary subsets of a group G for they do not reflect the fact that G has an algebraic structure imposed on it. Whatever subsets we do consider will be those endowed with algebraic properties

derived from those of G. The most natural such subsets are introduced in the DEFINITION: A nonempty subset H of a group G is said to be a subgroup of G if, under the product in G, H itself forms a group. (Herstein 1975: 37)

We learn "G has an algebraic structure imposed by it." Which is true of course, two skilled individuals are involved. Matchmaker1 built the Cartesian product of G × G creating all couples (a,b), and then matchmaker2 (the binary operation), following certain strict rules, created triples by matching each couple in G × G with another element of G. Group G is now a well-structured society, not just a shapeless set G, which only served as rough material from which the group was formed. And a subgroup must also be thus structured, must have the structure "imposed on it," or must be "endowed with algebraic properties derived from those of G." So no more busy hammering now; all is built and finished already; we just have to spot an independent and self-reliant substructure.

We mentioned above that to try to see the beauty of mathematics, we need the notion of a subgroup. So far, there is nothing particularly beguiling about it—a self-reliant, closed substructure of a larger structure, like a hospital wing, for example, part of the hospital but with its own power supply, entrances and exits, skilled staff, management, etc. Impressive perhaps, but not particularly beautiful. But then comes a series of lemmas that lead to the so-called Lagrange's theorem. And on the way to it, we learn about "right and left cosets." A very simple notion, a coset of H is Ha, that is, all the elements we get when we "multiply" (multiplication is just one example of a binary operation but let's stay with this simplification for a while) any x in H by an element a which is in G. Ha and aH are examples of right and left cosets of H, respectively. And we learn many things about cosets. For example, they are subsets of G of course because G is a group and therefore "closed" with respect to its binary operation. It was part of the definition or—in our story—part of matchmaker2 job description. Matchmaker2 was not allowed to associate ordered pairs in G with elements outside G. It is also proved that any two cosets are either identical or disjoint. Additionally, we learn that a union of the disjoint cosets of any subgroup H in G equals G, which means that those cosets "cover" the whole group. And also that "there is a one-to-one correspondence between any two right cosets of H in G" (Herstein 1975: 41). And one-to-one correspondence means that they have equal number of elements. But, since He = H (e is the identity element, like 1 in multiplication) is also a coset, all cosets have exactly o(H) elements and are "of the order of H," i.e., have the same number of elements as H.

Equipped with all those "facts" about subgroups and cosets, we are now just one step ahead of formulating the famous Lagrange's theorem, which is considered beautiful by mathematicians, and we will quote it in a moment, but the beauty is not only contained within a very simple and concise statement (which is coming) but the whole dramatic build up to it; all the features of subgroups and cosets we have just enumerated. And we will see it in just one moment, but first we must realize that beauty cannot be defined (even mathematicians know it). Beauty is a holistic impression—which sounds dangerously close to a definition. Let us not go there. Instead, let us use an example. In art galleries we rarely see people with their noses touching the paintings—and not only for security reasons but because we usually want to "take the whole painting/picture in." We can stay with each of the details separately, but then we want to see them all as a sum; we "look at the painting." And the quotation marks in the previous sentence are justified because by "looking" here we mean an "active search for beauty." And we are on the border of overusing quotation marks. Any moment now, if we do not want to get lost in the hopeless task of defining beauty, we will have to say "you know what I mean." And this is the moment, dear reader. If you ever enjoyed a painting or a sculpture in an art gallery, you know what I mean.[5] Up to now, in this non-finite paragraph, we have been looking at details (noses close to the canvas). It is time now to step back and take the whole painting in.

I will consciously use a parable now. A parable of looking through a kaleidoscope. As with any parable, we will try to understand one

---

[5] It seems so "unmathematical," doesn't it? How can we hope to say anything of value about mathematics, if we use phrases like this? But notice that mathematical narrative contains exactly the same phrase ("you know what I mean") in many places. And we have already seen three examples in Chaps. 3 and 4: the set, the element, and the ordered pair. They are never defined and we call them "primitives" or "primary notions." So anytime such undefined term is used, what we really have in the narrative of mathematics is something like: "and a set/element/ordered pair is.... erm... you know what I mean."

Table 5.1 The kaleidoscope parable

The kaleidoscope	Lagrange's theorem
A few bits of colored glass inside a tube	The symmetrical structure of group G
Look inside through symmetrical mirrors	Find a subgroup, H
Just turn a few times, left or right, by any angle—doesn't matter	Just multiply H a few times, from the left or from the right, by any element of G
To see perfect, stunning symmetrical shapes covering your whole visual field (Fig. 5.2)	To see equal, disjoint, symmetrical cosets covering the whole group

story through another. In our case, the story of Lagrange's theorem for finite groups through the story of looking through a kaleidoscope. In *The Literary Mind*, Mark Turner wrote that "parable serves as a laboratory where great things are condensed in a small space. To understand parable is to understand root capacities of the everyday mind" (1996: 16). Story and projection (mapping)—the basic ingredients of parable, "the root capacities of the everyday human mind"—have been with us at every stage of our analysis of mathematical narrative so far. And they will not leave us even when we dare to try to see the beauty of mathematics. So, finally, here it comes (Table 5.1).

We have not defined or proved anything. It was just another parable, a subjective impression of one reader of mathematical narrative. And we have not even seen Lagrange's theorem yet, so here it comes. But we may think of it as just an artist's signature on a finished work of art.

> If G is a finite group and H is a subgroup of G, then o(H) is a divisor of o(G). (Herstein 1975: 41)

So, for example, if G has ten elements, any subgroup H of G can have one, two, five, or ten elements, because only 1, 2, 5, and 10 are divisors of 10. But we knew it already of course; it is just a sum of details we enumerated above. Let's remember them again:

If H is a subgroup of G

- Each coset of H (Ha or aH) has the same number of elements as H.
- The cosets of H are either equal or disjoint.
- The cosets of H cover the whole group G.

What we have above is decomposition (division) of G into equal, disjoint parts, each of the order o(H). A kaleidoscope image known as Lagrange's theorem. And there the beauty dwells.

Fig. 5.2 A kaleidoscope image

Stunning as it already is, we have not seen all of it yet. Let us keep looking at the kaleidoscope image to see even more symmetries. But first we need one more simple notion—the so-called normal subgroup.

5.8 Normal Subgroups and the Beauty of Mathematics (Part 3)

DEFINITION A subgroup N of G is said to be a normal subgroup of G if for every $g \in G$ and $n \in N$, $gng^{-1} \in N$. (Herstein 1975: 50).

For example, for abelian groups (where the order of "multiplication"[6] does not matter), any subgroup is a normal subgroup. And normal subgroups have many exciting features, and one of them is that the product of any two right cosets is also a right coset. So we can say that

[6] For simplicity, we use the familiar term of "multiplication" to denote the group binary operation.

the set of right cosets is "closed" with respect to multiplication. Does it remind us of anything? Of course—matchmaker2. One of the rules she had to follow when "forming" a group was never to "leave" the group and never to associate outside the group. The group must be closed under its binary operation. It was part of the group definition. This is where the author of our (now definitely favorite) algebra handbook gets really excited and starts using exclamation marks. "The product of right cosets is a right coset. Can we use this product to make the collection of right cosets into a group? Indeed we can!" (Herstein 1975: 51). And then of course, a proof follows.

Herstein did not explain using the exclamation mark nor did he say anything about beauty (he mentioned it only once, many pages before, and only as a general reference to finite groups). But we know why he used it. Let us explain. What does it mean for our kaleidoscope image? We have just said that all those lovely cosets, which are equal and disjoint and cover the whole group, are also a group! And now I am excited and using exclamation marks. They also have the symmetry of a group. There is a second level of symmetry to the image. Not only is the group neatly and symmetrically divided into disjoint and equal cosets but those cosets have their own structure. They are also a group. A group of cosets. They are a group of equal and disjoint subsets of G. But this is not the end of it, because this new group, the "second level" group, has the same symmetry as G. So it can now be "covered" with its own "second level cosets." But of course, those "second level" cosets (cosets of cosets) are also a group. So there is now a third level of symmetry, etc. And it never ends. What we have here is "infinite symmetry." Well, almost—remember we are talking here of finite (and also abelian, to simplify the image) groups. So there is a finite number of symmetry levels. But finite can mean any number. So now try to imagine a kaleidoscope image with 2016 levels of symmetry, for example. And this is exactly why the exclamation mark was used by Herstein! And by me, in the previous sentence. And there even more beauty dwells. Beauty that is accessible to anyone, not only mathematicians. We just need to make a little effort and go the gallery or pick up a kaleidoscope.

Truth be told, we have not seen all the symmetries yet of the almost infinitely[7] intricate kaleidoscope image above. To see even more, we need another crucial mathematical notion—the homomorphism.

5.9 The Homomorphism

5.9.1 Homomorphism and the Carrier Story

> If there is one central idea which is common to all aspects of modern algebra it is the notion of homomorphism. By this one means a mapping from an algebraic system to a like algebraic system which preserves structure. We make this precise, for groups, in the next definition.
>
> DEFINITION A mapping φ from a group G into a group F is said to be a homomorphism if for all a,b $\in$ G, φ(ab) = φ(a) φ(b). (Herstein 1975: 54)

Homomorphism as we learn is a mapping. As we remember from Chap. 4, there are "many ways of thinking about" a mapping. Herstein "prefers to think of it" as a rule that associates elements—a match-maker. But we also saw that mapping can be understood parabolically as the story of "a carrier"—an actor who carries elements of sets "onto" other elements, for example, the "particular mapping which takes every real number onto its square" (Herstein 1975: 10). So let us use our freedom of choosing various "ways of thinking" about mappings and consider homomorphism as a carrier. The (typical for any definition of a mapping) use of the prepositions above—"A mapping φ *from* a group G *into* a group F"—certainly prompts for the choice of the scenario of carrying objects from one place into another. And the path along which objects are being carried is a vital part of this small spatial story which is then projected onto the story of a mathematical mapping.

---

[7] Finite, as we mentioned, can mean any number—however large—which of course begs the question of what non-finite is. But because we are trying not to jump ahead in this book, we will address this question later, when the mathematical narrative takes us to it. For now, let us use the following excellent definition by Douglass Adams. "Infinite: Bigger than the biggest thing ever and then some. Much bigger than that in fact, really amazingly immense, a totally stunning size, 'wow, that's big', time. Infinity is just so big that by comparison, bigness itself looks really titchy. Gigantic multiplied by colossal multiplied by staggeringly huge [...]."

But homomorphism is not just any mapping; it "preserves structure." And preserving structure, just like the use of the prepositions above, also prompts for "the carrier story." Let us concentrate for a moment on the concept of preserving structure. The first meaning in the Google Dictionary is "to maintain (something) in its original or existing state."[8] So we can imagine a millionaire buying a castle in Scotland to have it rebuilt on his estate in California or better on his other estate in Ohio, because it snows there from time to time, which will make the "new" castle look even prettier and more authentic. And the carrier/builder is employed to take down the Scottish castle brick by brick, carry it to Ohio, and rebuild it there as a perfect replica—preserving the original structure.

Let us take closer look at the "original structure" of the group. As we remember, two matchmakers were involved in the process of building (forming) of the group G. Matchmaker1 built the set of ordered pairs, the Cartesian square of G × G. And then matchmaker2 associated each of the pairs with another element of set G, following the four group definition rules. So the group is in fact organized in triples. For example, 2 + 3 = 5. The ordered pair of (2,3) is associated with number 5 (integers with addition are a group). And it's an ordered pair of ((2,3),5).[9] A pair where the first element is also a pair. And this is how the Scottish castle of G is built, and the Ohio castle of F must be built in exactly the same way. Let us call our particular "castle" homomorphism "OH" (like the Ohio state abbreviation). And imagine 2,3, and 5 are the bricks (or numbers of particular bricks) of the Scottish castle, and the bricks, when carried to Ohio and stamped by US customs, are now OH(2), OH(3), and OH(5). Our carrier builder must remember that brick 2 was first mortared to brick 3 and then the already mortared pair was mortared to brick 5. This was the original

[8] https://www.google.pl/webhp?sourceid=chrome-instant&ion=1&espv=2&ie=UTF-8#q=preserve%20meaing, accessed 2016-12-20.

[9] Following Herstein (and most other handbooks of mathematics), here is a task that the reader can solve. One of the ordered triples in the Scottish castle is ((1,1),2). But each number of course (number 1 included) indicates a *unique* brick. There is only one brick bearing number 1. *Task*: please explain why the Scottish castle is still standing strong, despite there being evidently a missing brick (the second number 1 brick). In fact, by analogy, many bricks are missing—consider ((2,2),4), ((3,3),6), etc. Hint: read the first part of Chap. 4 again.

structure. To preserve the structure and rebuild the castle as a replica, OH(2) must be mortared to OH(3), and then both bricks as a mortared pair must be mortared to OH(5). And, as we saw in the previous sentences, a lot of mortaring is required. So the Ohio triple building block must have this structure: ((OH(2),OH(3)),OH(5)). And this is exactly what the definition of homomorphism says because we can now write OH(2) + OH(3) = OH(5) or, to see it even more clearly, OH(2 + 3) = OH(2) + OH(3) ($\varphi(ab)=\varphi(a)\varphi(b)$).

We will now repeat the same tale of homomorphic castles using a matchmaker parable, which as we mentioned is Herstein's preferred way of thinking about a mapping.

5.9.2 Homomorphism and the Matchmaker Story

This tale is almost identical with the previous one but with one crucial difference: the other structure must be already there, built separately from Ohio bricks that were baked from native Ohio clay in a Cleveland brickyard. And what matchmaker5 (let's call her that, we'll explain in a moment) would be doing is simply checking whether all the bricks in Ohio were mortared correctly, preserving the original (still existing) Scottish structure, and whether indeed OH(2 + 3) = OH(2) + OH(3), which is just a notation for the following brick connections: ((OH(2),OH(3)),OH(5)). And we may now see why Herstein prefers to think of a mapping as a rule (a matchmaker) associating elements. It is because this particular parabolic projection works much better here. It simply matches the definition better because in the definition, F is already a group, a finished castle, which means that, as with any group, two Ohio matchmakers had to form it—matchmaker3 and matchmaker4. And they had to do this job before matchmaker5 (the number obvious now) could start checking the quality of the replica castle. Yes, homomorphism is just a building quality check. So we must be very careful which of the possible many "ways of thinking about" a mapping we choose if we want our small spatial story to match the definition of a homomorphism. And the fact that Herstein himself uses the word "preserve" and that we have the standard "from" and "into" in the definition does not make this task easier, because as we noticed earlier, those words prompt for the "carrier" story. And in the carrier story, the homomorphism (i.e., the carrier) starts working before the other castle is ready. In fact, this actor builds the new castle. And this is a mismatch, because in the definition, castle F, the Ohio castle, is already built (F is a group).

And this is exactly why mathematics is sometimes so difficult to learn. It is not just the complexity of it but the natural language used in the description ("preserve") and in the definitions ("from," "into"). Natural language which prompts us to pick up an incorrect small spatial story to project, which is of course vital because, as we are trying to convince the reader in every section of this book, small spatial stories and conceptual integration are crucial for understanding anything. If we pick the wrong story, we will understand homomorphism but just not the way mathematicians do. And they know which one to pick; Herstein told us (even if a touch obliquely and in a different chapter) of his preference for matchmaker (rule that associates elements). Unfortunately, not all algebra handbooks are as excellent as his. What we need in mathematical education is recognizing the crucial role of small spatial stories and conceptual integration for understanding and then applying this insight in the new handbooks.

Before we proceed to the summary of Chap. 5, let us take a final glimpse at the kaleidoscope beauty of finite groups.

5.9.3 *Homomorphism and the Beauty of Mathematics (Part 4)*

> LEMMA Suppose G is a group, N is a normal subgroup of G; define the mapping φ from G to G/N by $\varphi(x) = Nx$ for all $x \in G$. Then φ is a homomorphism of G onto G/N. (ibid., 56)

We learn that if N is a normal subgroup (and for abelian groups, any subgroup is normal as we mentioned above), group G is homomorphic with the set of right (and left of course) cosets G/N, which means that group G has the same structure as the set of equal and disjoint cosets of N that cover it. The group of cosets is a "replica castle" of G. So not only can we have 2016 (or more) levels of symmetry in our kaleidoscope image but each of those levels is a mirror (homomorphic) image of G. So we can have any number of "nested" symmetry levels but all those levels of symmetry have the structure of G.[10] And what was G again? A castle in Scotland of course. And in the kaleidoscope, this castle is covered with equal disjoint smaller castles (cosets of N). But the set of cosets is also a castle. And it is homomorphic with G. And this new (Ohio) castle is covered with equal disjoint smaller castles, and just repeat the process 2016 times to "get the picture"—the finished kaleidoscope image of Lagrange's theorem (Fig. 5.2).

Summary

In this chapter, we followed the mathematical narrative of the group theory. And of course, as in the previous chapters, we found the telltale traces of the way mathematical (i.e., human) mind works and creates. For example, we closely observed how two busy individuals (matchmaker1, matchmaker2) form a group (build a castle) out of the rough material of a set. We then followed all the consequences of it. And while following them, we did not reject "impressionistic reading and imprecise intuition" (Stockwell 2002: 6) and dared to say a few words about the beauty of mathematics or, more specifically, the beauty of finite groups expressed by Lagrange's theorem—one of the many gems of the group theory, each of them separately and uniquely beautiful. Not for a moment (not even when we tried to capture the elusive mathematical beauty) did we stray from our main subject, which is to see how small spatial stories and conceptual blending make mathematics meaningful and effective in modeling our world (Table 5.2).

As in Chap. 3, we came across the "one and many at the same time" problem. A fascinating feature that this time was explicitly addressed by the mathematical narrator. The identity element in a group was proven to be unique. But at the same time, it is not unique because, for example, the ordered pair of (e,e)[11] is an element of the group structure, which includes the Cartesian square of G × G. And yet, due to the (mostly subconscious) conceptual blending, this problem is unnoticed,

---

[10] In our kaleidoscope parable, the group structure of G was a projection of the kaleidoscope's symmetrical mirrors. And the kaleidoscope image always reflects their symmetry, just like the cosets reflect the symmetry of G.

[11] "e" is typically used for the group identity element (e.g., 1 for multiplication, 0 for addition, etc.)

Table 5.2 Elements of small spatial stories and traces of conceptual integration found in the narrative of groups

Objects	Group elements, ordered pairs, ordered triples, bricks, kaleidoscope
Actors	Various matchmakers creating the group structure, the truth collector (proof), builders, kaleidoscope user
Actions	Matching, mortaring bricks together, creating the Cartesian square and binary operation structure, collecting, exerting force, turning the kaleidoscope, recreating, preserving the structure, carrying, moving
Image schemas	Container, source-path-goal, collection, link, object, process, compulsion, resistance
Conceptual blending	"The one and many problem," the identity and inverse elements in a group—Multiple tokens of a unique object

or if it is, it is not considered a contradiction because in our "everyday" cognition, objects are one and many at the same time.

We also saw how the language of definitions and their description can be misleading, prompting the wrong understanding—making us choose the incorrect small spatial story to project. And mathematicians, the authors of the handbooks, always know which small spatial story should be used, but they usually do not let us know. And this is because story and conceptual integration are not yet considered relevant by mathematicians, which may be one of the reasons mathematics is considered such a difficult subject.

We will now follow the mathematical narrative into the ring theory. And a ring is a group with two binary operations. A typical example of a ring is the set of integers with addition and multiplication.

Chapter 6
Rings, Fields, and Vector Spaces

6.1 Overview

In this last stage of our exploration of mathematics, we will analyze three more algebraic structures of increasing complexity—rings, fields, and vector spaces. Herstein begins his chapter on rings in the following way:

> As we indicated in Chapter 2, there are certain algebraic systems which serve as the building blocks for the structures comprising the subject which is today called modern algebra. At this stage of the development we have learned something about one of these, namely groups. It is our purpose now to introduce and to study a second such, namely rings. The abstract concept of a group has its origins in the set of mappings, or permutations, of a set onto itself. In contrast, rings stem from another and more familiar source, the set of integers. We shall see that they are patterned after, and are generalizations of, the algebraic aspects of the ordinary integers. (1975: 120)

We learn that various mathematical concepts have their "origins." The rings, for example, are "patterned after" the algebraic features of "ordinary integers." And what Herstein is telling us explicitly now is that mathematics is constructed like a literary parable—by projecting one story onto another. We are to understand one story—the story of rings—through another, the tale of integers. This is exactly what "patterned after" means. Herstein continues in the same vein to tell us that "the analysis of rings will follow the pattern already laid out for groups. We shall require the appropriate analogs of homomorphism, normal subgroups, factor groups, etc." (ibid.). So the story of rings

J. Woźny, *How We Understand Mathematics*, Mathematics in Mind,
https://doi.org/10.1007/978-3-319-77688-0_6

will also be a projection, a mapping from the story of groups. In the next quotation, we are reminded of Herstein's view on what constitutes "good mathematics" (cf. Sect. 5.1).

> With the experience gained in our study of groups we shall be able to make the requisite definitions, intertwine them with meaningful theorems, and end up proving results which are both interesting and important about mathematical objects with which we have had long acquaintance. (ibid.)

"Meaningful" and "interesting" in this context mean connected with "mathematical objects with which we have had long acquaintance." The abstract axioms of rings, fields, and vector spaces must account for what we already know about integers, real numbers, and groups. The parabolic progression of mathematical narrative mirrors the basic pattern of human understanding, as explained by Mark Turner in *The Literary Mind*: "Parable is the root of the human mind—of thinking, knowing, acting, creating, and plausibly even of speaking" (1996: 168). And the basic patterns of "parabolic thinking" are small spatial stories, image schemas, and conceptual integration. Let us look for them in the mathematical narrative of rings, fields, and vector spaces.

6.2 Definition of a Ring, Small Spatial Story of Three Matchmakers

> DEFINITION A nonempty set R is said to be an associative ring if in R there are defined two operations, denoted by + and ·, respectively, such that for all a, b, c in R:
>
> 1. a + b is in R.
> 2. a + b = b + a.
> 3. (a + b) + c = a + (b + c).
> 4. There is an element 0 in R such that a + 0 = a (for every a in R).
> 5. There exists an element -a in R such that a + (−a) = 0.
> 6. a · b is in R.
> 7. a (b ·c) = (a· b)· c.
> 8. a (b + c) = a· b + a· c and (b + c) ·a = b ·a + c ·a (the two distributive laws).
>
> (Herstein 1975: 121)

And indeed, as Herstein mentioned above, we recognize all the familiar elements of a group definition, and we can see immediately that an associative ring is a group under the "+" binary operation (called "addition" henceforth). R is also an abelian group because we can add in any order. There is also a second binary operation (called "multiplication"—the quotation marks justified because the associative ring is a generalization of integers and their familiar arithmetic operations). And we can see that second operation—the multiplication—is not defined as a "proper" group operation because neither the inverse nor identity elements are required to exist in R. And the two operations are connected with distributive laws (point 8) which is also familiar and known from primary school as multiplying brackets.

As we mentioned above, the familiar stories of groups and integers are now to be projected parabolically on the new story of rings. And "familiar" is certainly the key word here. Mathematics is often described as "linear," i.e., you have to master all the previous bits before you can move on to the next ones. But if we think of it in the context of parabolic projection, there is nothing surprising about it. To understand one story through another, we have to be familiar with "another." So this "linearity" only strengthens our conviction that mathematics is a narrative constructed a set of parabolic projections.

Notice that this time, the nature of binary operations (the matchmakers, carriers, etc.) is not even mentioned. We are supposed to remember them from the story of groups. So let us remember them now. Matchmaker1 has to create a set of ordered pairs (a, b), where both a and b are in R. It is the Cartesian product (square) of R × R. And then matchmaker2 (also known as "addition") associates each of the pairs with a third element of R, following the group rules. But there is another matchmaker now—matchmaker3 (the "multiplication")—which also associates each ordered pair of R × R with a third element of R. Three cooperating matchmakers. But how exactly do those matchmakers cooperate?

6.3 The Structure of the Ring

6.3.1 How the Ring Matchmakers Cooperate

How is the ring of R organized? For example, does matchmaker1 have to work overtime now and prepare two separate Cartesian squares for each of the other matchmakers, or will one suffice? If the second, if only one R × R is created, perhaps, we can think of the ring structure as the following triples (or quadruples) now ((a,b),a + b,ab) or maybe ((a,b), ab, a + b)? And the answer is a resounding NO. What we mean is we can think of the ring structure anyway we like, but there is only one "proper" ("mathematical") way allowed here. So, if we want to think of it the way Herstein and other brilliant mathematicians do, we have to follow the parabolic level-by-level progression of mathematical narrative very closely. We mustn't forget any details of the background stories. So let us go back to the ground floor again (we can use the elevator later to return to the ring level). Matchmaker2 and matchmaker3 are two separate binary operations (so the definition above says two). And what is a binary operation? A mapping. And what is a mapping? Well—as Herstein mentioned before—there are many ways of thinking about a mapping. And of course one of the "allowed ways of thinking" is the so-called graph definition we discussed in Chap. 1. A mapping is a set of ordered pairs. And what is a set and an ordered pair? Erm, uh, hmm... And this is where mathematical narrative will not help us anymore. Neither the set (also an element) nor an ordered pair is defined (cf. Ftnt. 14 in Chap. 2). They are the so-called "primitive" or "primary" concepts (we could add "primordial," "ancient," "venerable," and "hallowed" here, but we would still be none the wiser—and neither are the brilliant mathematicians). This is where mathematics says "you know what I mean",[1] dear reader, as we did when discussing the concept of mathematical beauty in the previous chapter. We are to rely on our own mettle—on our intuition and experience with collections and ordered pairs of objects. We went too far down now—this the dark, dusty cellar we mentioned in Chap. 3. Let's go back up to the lobby again, where there is light and air-conditioning.

And in the lobby we have two binary operations which are two mappings, two sets of ordered pairs, and each pair has the following structure ((a,b),c). So each of those pairs has a pair as the first element.

---

[1] One could think that this is where mathematics is weak and fuzzy, a sort of Achilles heel of mathematics. Conversely, I think this is exactly where the strength of mathematics lies. The undefined primitives give mathematics (via small spatial stories and blending) its amazing flexibility and effectiveness. Just as the fuzziness of human categories makes them so much more effective and energy-efficient in communication than the Aristotelian categories (cf. Rosch 1978).

Fig. 6.1 Intersecting binary operations in a ring

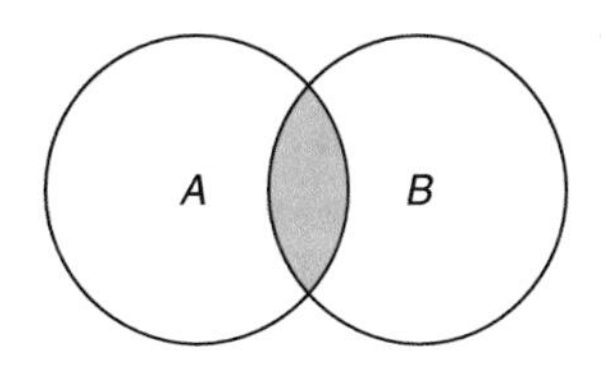

And of course a, b, and c are in R (are elements of set R). Two sets. And if we want to imagine them, the best way is to use those oval/circular shapes—Venn's diagrams. And there are only three versions of those diagrams if we have two sets. The three versions are separate shapes, not quite separate (intersecting), or one inside the other. So we have all three of them on the I-max 3D Dolby-digital screen of our imagination. Let us choose which of them should our camera focus on (it is just panning/trucking from left to right at the moment but ready to dolly forward and zoom on one of the three images at any time). Let us think. Does it ever happen that matchmaker2 and matchmaker3 (addition, multiplication) perform the same matches? And the answer is yes—possibly. For example, $2 \times 2 = 2 + 2$, $0 \times 0 = 0 + 0$.[2] So the image to focus on is this (Fig. 6.1):

Where A is the addition, B is the multiplication, and in the intersection area, we have, for example, ((2,2),4) and ((0,0),0). And this is how the two matchmakers, binary operations (addition, multiplication) are connected. Sometimes they may find the same match for the same couple (and then of course, the couple in question will demand a refund from one of them[3]).

And this is the story of rings so far—three busy matchmakers instead of two as it was with groups. In fact, the diagram above does not tell all about the connection of the two binary operations—matchmaker2 and matchmaker3. Axiom 8 of the ring definition

---

[2] A task for the reader: prove that those are the only integer solutions of $ab = a + b$.

[3] Well, the refund will probably not be granted because (2,2) and (0,0) are not "real couples" but just ordered pairs of one unique element (cf. our remarks on uniqueness in Chaps. 3 and 5). "I am really sorry, Sir, but Wholesome, Threesome & co. is an agency for couples, not just one person pretending to be a couple." On the other hand, 0 may get this refund anyway. It pretended to be a couple of (0,0), true, but was matched with itself by both agencies. Not fair really, because the "fake couple" of (2,2), for example, was matched with a delightful and quite separate character called "4."

(multiplying brackets) says, for example, that a(b + c) = ab+bc, which means that the binary operations, the matchmakers, do not work independently but have to watch one another closely. For example, let us imagine multiplication (matchmaker3) trying to find a match for the couple of (a, b + c). There is no more freedom now of just calling anyone. There is only one choice. The multiplication can only associate this couple with the match the addition found for the pair of (ab,bc), which is of course ab+ac. But it works the other way round too, since the equality relation expressed by "=" is symmetric. So whenever the addition (matchmaker2) is wondering what to do with the couple of (ab,ac), the only choice is the match already found by the multiplication for (a,b + c), i.e., a(b + c). So we have three cooperating matchmakers, two of them (matchmakers 2 and 3) partially intersecting (joined at the hip).

6.3.2 *Ring as a Closed Container, Force Dynamics of Proof*

Herstein comments on the ring axioms (listed at the beginning of Sect. 6.2) in the following way:

> Axioms 1 through 5 merely state that R is an abelian group under the operation +, which we call addition. Axioms 6 and 7 insist that R be closed under an associative operation ·, which we call multiplication. Axiom 8 serves to interrelate the two operations of R. Whenever we speak of ring it will be understood we mean associative ring. Nonassociative rings, that is, those in which axiom 7 may fail to hold, do occur in mathematics and are studied, but we shall have no occasion to consider them. (Herstein 1975: 121)

We can see in the above that the axioms are also actors capable of performing multiple actions—they "serve to interrelate," "insist that," "merely state," and sometimes even "may fail to hold." Axioms speak with a variety of voices. It all depends on their mood on a particular day. Sometimes it is quiet monotone, when they just indicate things, "merely state." But on other occasions, when they are in a more imperative mood, they will (we can imagine) raise their voice and "insist" categorically that the matchmakers, busy forming the ring structure, behave in a certain way. And as actors, the axioms not only speak but also perform physical actions. Actions that require the use of force, as in holding objects. We learn, however, that the axioms,

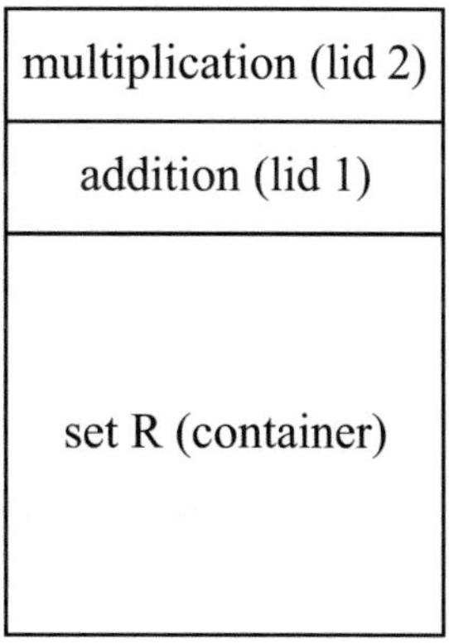

Fig. 6.2 Ring as a closed container

multiplication (lid 2)

addition (lid 1)

set R (container)

just like us, humans, have limited capacity of exerting force. And sometimes, try as they may, they "fail to hold."

"Axioms 6 and 7 insist that R be closed under an associative operation." R is a container then and the binary operation is a lid that is put on top of it to close it. And it's all done under the watchful eye of the axioms, which insist that the container of R be closed. So instead of the Venn's diagram above, we may now imagine the ring structure in the following way (Fig. 6.2).

We also learn that nonassociative rings "do occur in mathematics and are studied." And, as we noticed in Chap. 5, this is indicative of imagining mathematics to be external, independent of human cognition. Mathematical objects are out there, and we are just studying them—not as creators but merely as observers. And the next of Herstein's comments on the definition of an associative ring seems to confirm it. "It may very well happen, or not happen, that there is an element 1 in R such that $a \cdot 1 = 1 \cdot a = a$ for every a in R; if there is such we shall describe R as a ring with unit element" (ibid.). Mathematician is a botanist, discovering and collecting various, naturally occurring plants. "Natural examples exist where $a \cdot b \neq b \cdot a$. All these run counter to our experience heretofore" (Herstein 1975: 125).

So far, following Herstein's guidance of course, we have been using the familiar names of "addition" and "multiplication" for the ring binary operations. But what about division? We certainly know how to divide integers, for example, but the result (3/4 or 7/5, etc.) does not have to be an integer. We need a new structure, which will be "closed under division."

6.4 Rings, Fields, and Arithmetic

Heretofore in this chapter, we discussed the definition of a ring, and we are just a step away of another crucial mathematical construct (or natural species, as Herstein and many others would have it)—the field.

Definition A field is a commutative division ring (Herstein 1975: 126),

Where "commutative" means that for multiplication ab = ba and "division" means that all the elements but 0 (the addition identity element) form a group under multiplication. And the fact that the story of rings (and fields now) is a parable of (among others) rational numbers helps us immediately to understand why 0 has to be excluded from the multiplicative ring. If we went back to the group definition in Sect. 5.2, we would remember that there would have to exist an inverse element 0^{-1} such that $0{\cdot}0^{-1} = 1$. And of course no such rational number exists as multiplication by 0 always yields 0. A field then has a "double-group" structure—it is a group under addition and under multiplication. But it is also a ring, which means that the two binary operations are connected in the way we discussed above (the distributive laws—multiplying brackets). Of course, the set of rational numbers is just one example, and we expect, as is the mathematical way, a more general theorem and a formal proof. Herstein, as we remember from the first section of this chapter, told us explicitly how a good mathematical story should be written: "make the requisite definitions, intertwine them with meaningful theorems, and end up proving results which are both interesting and important about mathematical objects with which we have had long acquaintance" (Herstein 1975: 120). And this is precisely what he does next.

LEMMA 3.2.1 If R is a ring, then for all a, b ∈ R:

1. a0 = 0a = 0.
2. a (− b) = (−a) b = − (ab).
3. (−a)(−b) = ab.

If, in addition, R has a unit element 1, then:

4. (−l)a = −a.
5. (−1) (−1) = 1. (Herstein 1975: 126)

The above is certainly meaningful and of long acquaintance to anyone who knows basic arithmetic. When we first learned it, it probably never occurred to us that it can be proven. Why do we need to prove something we have known since primary school?

6.5 From Set and Element to Arithmetic: The Story So Far

The answer of course is that the familiar story of numbers (integers, real, rational, irrational, numbers, etc.) can now to be projected onto an infinite variety of all kinds of rings and fields. And not only that, mathematicians seem to be constantly to be on the lookout for the simplest and the most basic source stories. The source story here is of course, looking back at the foundation, set and element. This is how our algebra narrative began. The progressions, the narrative projection stages that we have been analyzing so far, are these: set/element/ ordered pair >> mapping >> group >> ring >> field. What is amazing here is that we are about to witness the proof of the most basic rules of arithmetic. The proof is ultimately based on the "primary" concepts of set and element. So what mathematical narrative is telling us is that, for example, $(-a)(-b) = ab$ is a projection of the story of collections—sets and elements. And, if anyone was wondering what mathematical beauty is, here is the answer again. Or a significant part of it.

But how is it possible to get from a collection of objects to $0a = 0$? It is nothing short of a miracle—one might think. But we have seen in this and the previous chapters exactly how it happened. Let us quickly summarize the story so far. It started with a collection of objects (elements in a set). And then a mapping was defined. And it was defined as a set of ordered pairs. But ordered pairs are also mappings. So the definition is circular and as such does not define anything. Fortunately, we were told that there are many ways of "thinking about" a mapping. And one of those ways was a story of a matchmaker that associates elements of sets. And this is how our story progressed. To create a group, two matchmakers had to work on a set, and then one more was needed to create a ring. The first of the matchmakers has to create a Cartesian square of RxR, which contains ordered pairs (a,b) of elements in R. And then the other two matchmakers associate those pairs with elements of R. Doing that, the matchmakers (addition, multiplication) have to follow a set of rules—the ring axioms (which are basically the group axioms). And this is it. This is where we are now.

So where is the arithmetic in this story? Well, let us look again at, for example, $0a = 0$. "0a" is the notation for the element matched by multiplication (matchmaker3) to the pair of (0,a). And 0 is the identity element of addition (matchmaker2). There is such element and it is "unique"[4] in R. So 0a = 0 is a shortcut for saying: matchmaker3 has to associate the pair of (0,a) with 0. It is a statement about the way one of the matchmakers that create the ring has to work. And this is what arithmetic is—rules for, or statements about, the matchmaking associations. 1 + 2 = 3 means that addition (matchmaker2) has to associate the pair of (1,2) with 3. And that's all there is to it. So this is how we got from sets and elements to arithmetic. There should be nothing mysterious about it—just a couple of matchmakers working to rule (like, e.g., Republic Act 6955[5]).

6.6 Multiplication by Zero Equals Zero: Proof as an Actor

Let us now enjoy part of the proof of LEMMA 3.2.1 above. "If $a \in R$, then $a0 = a(0 + 0) = a0 + a0$ (using the right distributive law), and since R is a group under addition, this equation implies that $a0 = 0$" (Herstein 1975: 126). This was fast, wasn't it? And now, our matchmaker3 (multiplication) knows that the ring axioms (its job description) bound it to associate every (a,0) couple with 0. And this is also a rule of arithmetic. And the other proofs, which we need not quote here, are equally short and simple, following directly from the ring axioms. Or, if we remember our remarks on axioms also being actors, this is what the axioms "insist on." The axioms are the matchmaker police; they watch the matchmakers closely and demand the rule of law be maintained. At least as long as they can uphold it. Unfortunately, as we learned, sometimes they "fail to hold," and then of course the matchmakers run riot and do what they want. Herstein summarizes the series of proofs of LEMMA 3.2.1 as follows. "With this lemma out of the way we shall, from now on, feel free to compute with negatives and 0 as we always have in the past. The result of Lemma 3.2.1 is our permit to do so. For convenience, $a + (-b)$ will be written $a - b$" (Herstein 1975: 127). We are reminded of our remarks in the previous

[4] cf. our remarks on uniqueness in Chaps. 3 and 5.

[5] Republic Act 6955 prohibits the business of organizing or facilitating marriages between Filipinas and foreign men (https://en.wikipedia.org/wiki/Online_dating_service, accessed 2016-12-28).

chapters on another actor—mathematical proof—also known as the truth collector. This character acted by moving along a path (the path of proof) putting objects out of the way.

6.7 Small Spatial Stories of Addition and Subtraction

6.7.1 The Small Spatial Story of Jenga Blocks

We learn from the above quotation where subtraction comes from—at least in the narrative of modern algebra. It was in fact already part of the story of groups, where the inverse element was introduced. Subtracting is simply adding the inverse element.

We could use the word "simply" in the previous sentence because we have been closely following the story of sets, mappings, groups, rings, and fields. But we can imagine that for a school student, when they first learn that $a - b = a + (-b)$, it need not be simple. Taking away is not adding, is it? And even the inverse element is not something that could easily be translated into our everyday experience, like collections or matching one object with another. What happens with the inverse element is that you have an object (number) and then add something to it and it disappears. The result is zero, nothing. Adding objects, say piling Jenga blocks one on top of another and then taking them away from the pile, is the way most of us understand adding and subtracting (Fig. 6.3).

And adding and removing (taking away), just like matching, can be found in most inventories of image schemas (e.g., Johnson 1987, Lakoff and Turner 1989). But it never happens, does it, that when we put one block on top of another they both disappear.

And it may seem the above magic trick of disappearing objects clashes with Herstein insisting repeatedly that mathematical axioms yield results that are "meaningful and familiar." Well, mathematics can be thought of as a narrative, a literary creation structured like a parable /blend (projecting/mapping stories onto other stories). So perhaps we should expect a Lord of the Rings magic component now and again. Well, perhaps, but we have not had any magic tricks so far. Our story could easily be tied down to small spatial stories and image schemas of collections of objects/containers (sets and elements),

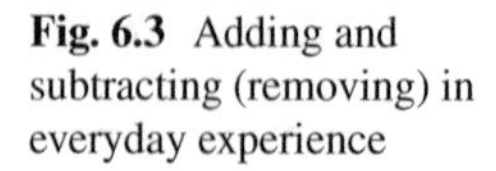

Fig. 6.3 Adding and subtracting (removing) in everyday experience

matching (mapping), and force/motion (axioms, proof). And I will try to demonstrate that this image-schematic base of our mathematical narrative still holds. And it holds under one condition only: that we follow the narrative closely. A magician's act can only be successful if we are not allowed to see all the moves—otherwise the magic disappears. So let us look even closer at the magician's act. And for that we will have to return for a moment to the story of groups.

6.7.2 Cayley's Theorem

In this volume we follow the story of "pure mathematics" (algebra in our case) as it is told today in modern handbooks, of which Herstein's (1975) *Topics in Algebra* is an excellent example. However, although Herstein's focus is the state of the art as it is today, sometimes, when it feels appropriate, he also mentions the historical roots of certain concepts.

Very early in his book, in Chap. 1, after defining a mapping, Herstein mentions the set of all one-to-one mappings of a set S onto itself—A(S). And comments on it in the following way: "Aside from its own intrinsic interest A(S) plays a central and universal [...] role in considering the mathematical system known as a group" (15). And then in Chap. 2, we find the following comment:

> When groups first arose in mathematics they usually came from some specific source and in some very concrete form. Very often it was in the form of a set of transformations of some particular mathematical object. In fact, most finite groups appeared as groups of permutations, that is, as subgroups 'of Sn. (Sn = A(S) when S is a finite set with n elements.) The English mathematician Cayley first noted that every group could be realized as a group of A(S) for some S. (Herstein 1975: 71)

Arthur Cayley, A Lucasian[6] professor at Cambridge University, was one of the founders of what we call today "modern school of pure mathematics." And that means he helped to bring mathematics from the domain of "concrete and specific" into the realm of "purely abstract." And in 1854 he formulated the theorem, mentioned above, which was later named in his honor. And "the theorem enables us to exhibit any abstract group as a more concrete object, namely, as a group of mappings" (Herstein 1975: 72).

The binary operation for this "concrete" group of A(S) is the "composition," also called the "product" of mappings. A very simple concept defined for every x in S as fg(x) = g(f(x)). So the product fg works in the following way. First x is taken onto f(x) and then f(x) is taken onto g(f(x)). And, we have deliberately used one of the allowed ways of "thinking about mappings" we discussed in Chap. 1. We used the story of a carrier. And this actor's job is very simple—to carry an object (a set S of objects) from one place to another. And the inverse element of f in A(S) is the inverse mapping, which is of course every carrier's nightmare and means that she now has to undo everything she has done and bring all the objects, elements of the set S, back to their original places, which in mathematical notation is written as $ff^{-1}(x) = x$ (for every x in S). So the product of f and its inverse element yields the identity element of A(S) which is the identity mapping. And the identity mapping is the job every carrier loves, because it means doing nothing and still getting paid. In the daily report, the

[6] The so-called Lucasian Chair of Mathematics is considered to be one of the most prestigious professorships in the world. Its occupiers over the years were, among others, Sir Isaac Newton, Paul Dirac, and—most recently—Stephen Hawking.

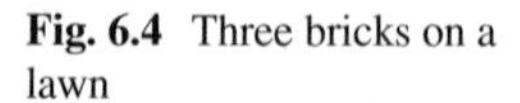

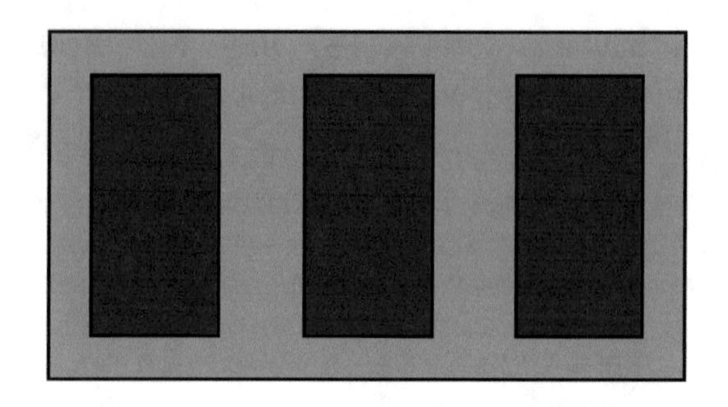

Fig. 6.4 Three bricks on a lawn

carrier would simply state "I brought every x onto itself" and then go and have her tea until it is 5 o'clock. So this is what a group originally was—a set of a carrier's jobs.

6.7.3 *The Small Spatial Story of Three Bricks*

To imagine it even better, and to bring the concept of one-to-one mappings of S onto itself—A(S)—more to light, let us start with imagining S first. S is a set. And what is a set? A collection of objects. So let us imagine a concrete collection. Say, three bricks. This is an important imagination exercise, dear reader; please bear with me. Three separate bricks. And of course we cannot have them hanging in the air, so let us imagine three red bricks on a green lawn on a lovely summer's day. And they are lying flat on the grass next to one another, 10 cm apart, forming a neat rectangle (Fig. 6.4).

This is how we imagine collections of objects. Objects, as we know them, occupy space. They can move of course but let our collection be a static one. And as we look at our three red bricks on a green lawn, we already have them ordered: left, middle, and right; or west, middle, and east; or 1,2, and 3; or right, middle, and left; etc. Yes, ours is an orderly collection. And, as it is a lovely summer's day, and we have nothing better to do, we can now play with those three bricks. For example, put the right one in the middle and the middle on the right. We are the carrier now. What we just did was a mapping. This is exactly what a mapping is—carrying objects from one place to another. We can now reverse what we just did and put the middle back on the right and the right one back in the middle. This is fun, isn't it?

What could be nicer than playing with three bricks on a lawn on lovely summer's day. We have just applied an inverse mapping to our set S of red bricks. So this is what an inverse mapping (or an inverse element in a group) is—taking the bricks back to their original position. And we can play with bricks all day of course. One of the many fulfilling ways to play with them could be, for example, to lift each of them a little and put it back again where it was, without changing places. Yes, we carried the bricks but left them where they were. And this was also a mapping, a very special kind of mapping—the identity mapping (or the identity element in a group). When we changed the position of the right one and the middle one and then put them back where they were, the result was an identity mapping. Our brick-moves, acts of carrying bricks from one place to another are a group. This is the origin of the concept. This is where mathematical groups (finite and non-finite) come from. We can understand the story of groups through the story of bricks. So let us now go back to the group definition from Sect. 5.2 but look at it as a parable of our three bricks on a lawn.

DEFINITION A nonempty set of elements G is said to form a group if in G, there is defined a binary operation, called the product and denoted by·, such that (Herstein 1975: 28):

1. $a, b \in G$ implies that $a \cdot b \in G$ (closed).
2. $a, b, c \in G$ implies that $a \cdot (b \cdot c) = (a \cdot b) \cdot c$ (associative law).
3. There exists an element $e \in G$ such that $a \cdot e = e \cdot a = a$ for all $a \in G$ (the existence of an identity element in G).
4. For every $a \in G$, there exists an element $a^{-1} \in G$ such that $a \cdot a^{-1} = a^{-1} \cdot a = e$ (the existence of inverses in G).

G is now a set of carrier jobs, a set of moves, all the ways we can play with our bricks by changing their positions.[7] And when we do two consecutive brick-moves, what we get is also a brick-move. Are bricks moves associative? Yes they are. It can easily be proved but let us just have a brick example. Let us play with our bricks some more. Imagine the following three moves: move 1 (right, middle), move 2 (middle, right), and move 3 (right, middle). So what we did was put the right brick in the middle and the middle brick on the right. Then, in move 2, we reversed it. And then, in move 3, we repeated move 1 again. And as a result, the right and the middle brick changed

[7] A task for the reader: prove that there are $1 \times 2 \times 3 = 6$ possible "brick-changing-places moves."

positions. And now let us do the same moves but in different order: move 2 (middle, right), move 3 (right, middle), and move 1 (right, middle). And, of course, the result is the same. This is what "associative law" means. Is there an identity element in our brick-moves? Yes there is. It's when we do nothing. Is there an inverse brick-move for every brick-move? Yes there is—we just have to put the bricks we moved back to their original places.

We can see now how the story of playing with bricks (brick-moves, also referred to as "permutations") was projected onto the story of groups. And then the story of groups was projected onto the story of rings and fields like, say, real numbers with addition and multiplication.

6.7.4 From Bricks to Arithmetic

Let us concentrate on addition (subtraction) because this was the reason for our detour back into the group (brick) theory. Adding two numbers means performing two brick-moves. $1 + 2 = 3$, for example, means that when we perform brick-move 1 and then brick-move 2, what we get is brick-move 3. $2 + (-2) = 0$ means that when we perform brick-move 2 and then put the bricks back as they were again, what we get is doing nothing, i.e., the identity element, a zero-move. And we already see what negative numbers are as brick-moves—they are the reverse moves, the go-back moves. And where's the magic of disappearing Jenga blocks? Gone in a puff of smoke (or brick dust). We can now understand both subtraction and negative numbers easily through brick-moves. Take $4 + (-6) = 4–6 = -2$. What does it mean in the language of moving bricks? Easy: perform move 4 and then the reverse of move 6, and what you get is the reverse of move 2.

And again, the adjective "easy" was used in the previous paragraph because we have seen the connection of "brick-moves" (group permutations) to numbers. A connection definitely present in the narrative of modern algebra, where we are prompted, for example, to see real numbers as a field, which is a ring, which is a group. And by force of Cayley's theorem, any group is a group of permutations (brick-moves) of a certain set, which is a collection of objects that can be carried (mapped) "on themselves." But will it help our school pupil to make sense of $4 + (-6) = -2$? And the answer is no it won't. She will never learn of Cayley's theorem until she is in her second year at Cornell University, enjoying Herstein's advanced algebra handbook. And even then, she might easily miss the significance of Cayley's theorem. Even in this book, where we try to follow the narrative of algebra as closely as possible and without skipping ahead, we almost missed it. It was only quoted as an afterthought, when we puzzled over disappearing Jenga blocks.

Should we perhaps add Cayley's theorem to primary school curriculum? Certainly not. Let us leave it where it is part of an advanced university course. But we could definitely use the idea (implicit in Cayley's theorem and in the "origin" of the group concept) of addition/subtraction being completely equivalent as a product of mappings (permutations), which themselves are acts or carrying objects (e.g., bricks). What could be simpler in this context than $1 + (-1) = 0$? It just means taking (carrying) a brick to a different place and bringing it back again. The story of brick-moves. And as a result, the brick is where it was and this is what 0 means. And in this way, we would apply the core of advanced algebra (Cayley's theorem) in primary school teaching. And we could do that because ever since *Where Mathematics Comes From*[8] (Lakoff and Nunez 2000), we realized that "cognitive debunking," demystifying, mathematics is possible.

Yes, mathematics can be demystified and yet stay full of its natural (as opposed to supernatural, magical) beauty. The beauty residing in, for example, the kaleidoscope symmetry of finite groups (the Lagrange's theorem we discussed in Chap. 5) or just the simplicity of the source stories (collections, carrying/matching) which are projected (as inputs in conceptual integration networks) on the infinite number of increasingly complex structures (like groups, rings, and fields) in the parabolic narrative of algebra. And when we use the phrase "natural," we don't mean—as Herstein often implies—that mathematics is "out there," independent of the human mind. On the contrary, the beauty is natural because it relies on the natural features of human cognition, such as small spatial stories, projection, image schemas, and conceptual integration.

[8]The full title is of course *Where Mathematics Come From: How The Embodied Mind Brings Mathematics Into Being*.

6.7.5 Arithmetic at School vs. Modern Algebra

The mathematics teaching methodology is not the primary focus of this volume, but let us say a few more words on the possible practical application of our "debunking" of algebra for teaching the concepts of addition, negative numbers, subtraction, and zero. They should, in our opinion, all be introduced to pupils at the same time (as a complete set) because all four of them, as we found above, are be based on the primary concept of motion and carrying objects. -2 is just as natural as 2—it is just a reverse move. And when we reverse a move, we get back to the starting point—and this is what zero is. And what we have just said is not just another random way of explaining the mysterious mathematical concepts. This is exactly what all group binary operations (mappings) are. This is the (brick) origin of groups (and, consequently, rings and fields) we found by exploring the narrative of advanced algebra.

The traditional method of teaching arithmetic is to start with addition and then gradually introduce the "more complex" concepts of zero, subtraction, and negative numbers. The following examples come from a handbook for teachers called *Practical Approaches to Developing Mental Maths Strategies for Addition and Subtraction*[9] based on the insights of a well-known mathematics educator—John Van de Walle[10]—and his 6th edition of *Elementary and Middle School Mathematics Teaching Developmentally* (Van de Walle 2007). The book contains the relevant quotations followed by practical tips for teachers and exercises that can be used in the classroom.

> Occasionally pupils feel that 6 + 0 must be more than 6 because 'adding makes numbers bigger' or that 12–0 must be 11 because 'subtracting makes numbers smaller'. Instead of making arbitrary-sounding rules about adding and subtracting zero, build opportunities for discussing zero into the problem-solving routine. (Van de Walle 2007: 154)

The ensuing practical advice for the math teacher is to "pose problems involving zero. For example, Robert had eaten 8 grapes. He was

---

[9] http://www.pdst.ie/sites/default/files/Mental%20Maths%20Workshop%201%20Handbook.pdf, accessed 2017-01-03.

[10] https://www.nctm.org/Grants-and-Awards/Supporters/John-A_-Van-de-Walle-Biography/, accessed 2017-01-04.

too full to eat any more. How many did Robert eat altogether? In discussion of the problem, use drawings/counters to illustrate the empty set (zero)." The "unnatural and complex" concept of zero is to be taught separately from addition/subtraction, with the use of a rich variety of teaching aids and completely redundant questions about Robert. Yes, dear children, he has eaten 8 grapes—how many grapes has he eaten?

Subtraction, according to Van de Walle, must also be tackled with care and only after the pupils have got the hang of addition.

> Evidence suggests that children learn very few, if any, subtraction facts without first mastering the corresponding addition facts. In other words, mastery of 3 + 5 can be thought of as prerequisite knowledge for learning the facts 8–3 and 8–5. Without opportunities to learn and use reasoning strategies, students may continue to rely on counting strategies to come up with subtraction facts, a slow and often inaccurate approach. When children see 9–4, you want them to think spontaneously. 'Four and what makes nine?' (Van de Walle 2007: 175)

To heed Van de Walle's guidance, teachers can use, for example, the following exercise.

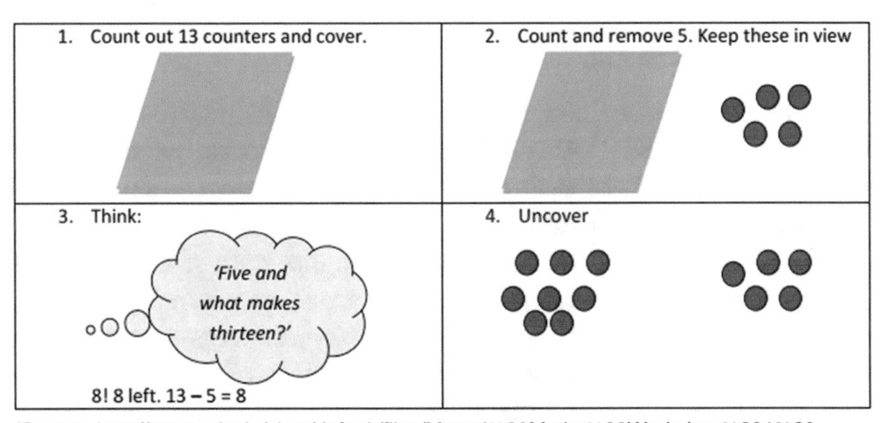

(Source: http://www.pdst.ie/sites/default/files/Mental%20Maths%20Workshop%201%20Handbook.pdf, accessed 2017-01-03)

It is surprising, isn't it, how much what we teach at elementary level differs from the simple elegance of advanced algebra—Cayley's theorem, for example. It feels like a completely different subject and much more complicated at that. And by strength (!) of Cayley's theorem, let us restate, there is nothing mysterious or complicated about adding the inverse element (subtraction)—it is just putting the bricks back where they were. And what results is a zero-move, i.e., bricks stay in their original place.

As we noticed before, part the beauty of modern algebra narrative is the simplicity of its source stories (e.g., carrying objects from one place to another). But, unfortunately, all this brilliant elegance and simplicity has yet to find its way into the elementary classroom.[11] And I hope that this book, and cognitive exploration of mathematics, in general, can help to achieve that.

Let us terminate our short visit to the primary classroom and return to the narrative of advanced algebra. In the next section, we will introduce the most intricate algebraic structure so far—the vector space.

6.8 Vector Space and the Seven Matchmakers

This is how Herstein begins his chapter on vector spaces:

> Up to this point we have been introduced to groups and to rings; the former has its motivation in the set of one-to-one mappings of a set onto itself, the latter, in the set of integers. The third algebraic model which we are about to consider-vector space-can, in large part, trace its origins to topics in geometry and physics. (1975: 170)

As in the previous chapters, we learn that abstract mathematical concepts have their "concrete" origins. In this case, geometry and physics. And it is not the first time that physical objects will be used as a source story in mathematical parable. As we remember from Chap. 3, the story of modern algebra begins with a collection of objects. Yes, physical objects in physical space, so geometry was also already involved. In fact, geometry and physics accompanied us on every stage of mathematical narrative. The story of a collection progressed into the story of a mapping. And one of the ways a mapping can be interpreted is carrying (physical) objects (from a collection) from one (geometric) place to another. Further on we learn that the

[11] Herstein seems to be aware of this disparity when he says, for example, "Very early in our mathematical education—in fact in junior high school or early in high school itself—we are introduced to polynomials. For a seemingly endless amount of time we are drilled, to the point of utter boredom, in factoring them, multiplying them, dividing them, simplifying them. Facility in factoring a quadratic becomes confused with genuine mathematical talent" (Herstein 1975: 153).

origin of a group is a set of physical moves (the bricks on a lawn from the previous chapter), i.e., a set of permutations. The mappings that are referred to as "binary operations" are compositions of physical moves. And the physical moves (mappings, permutations) are composed in the following way: a physical object is taken from A to B and then from B to C. The resulting move (the composition or multiplication of mappings) is of course the move from A to C. If certain conditions are met, the set of physical moves becomes a ring or even a field as we saw in the previous chapter.

The composition of moves we described above could be written, for example, as (A,B) + (B,C) = (A,C). And we are already adding vectors. So we are pretty well there and could skip to the next chapter, but we will not because we promised the reader not to skip ahead but instead to follow the narrative of algebra as closely as it is possible in this short volume. So let us return to our "Topics in Algebra" where we find next that:

> Vector spaces owe their importance to the fact that so many models arising in the solutions of specific problems turn out to be vector spaces. For this reason the basic concepts introduced in them have a certain universality and are ones we encounter, and keep encountering, in so many diverse contexts. (Herstein 1975: 170)

Vector spaces seem to pop up everywhere, but perhaps it should not be surprising in the light of our remarks above—they are already embedded in the narrative of algebra even before they are defined. And we can see that because we are able to pin down the "mental patterns of parable" (Turner 1996) in the narrative of mathematics. Those mental patterns always include objects, actors, action, projection, image schemas, and conceptual integration. This is exactly how we know that binary operations that define groups, rings, fields, and (in a moment) vector spaces are based on a story of carrying objects along a path from one place to another. Those objects are sometimes returned to their original place (this is what the so-called inverse element in a group does) and what results is a no-move, a zero-move (also known as the identity element).

We do not know the story of vectors—it hasn't started yet—but what we have seen so far is mathematical (algebra) narrative where the same small spatial stories appear again and again. And one such story, which seems to be crucial so far, is the story of objects being moved along a path—the story of mappings. Let us draw it schematically.

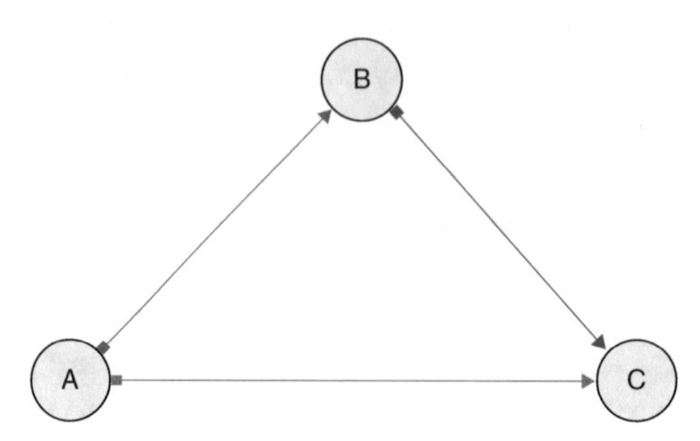

An object is moved from A to B and then from B to C. This is one of the source stories for mappings and their "product" or "composition." It does not matter where points A, B, or C are. Whatever their position and however close or far away they are, the result is also a move (a mapping), a move from A to C. And this feature of moving objects from place to place was projected onto the story of groups. A group must be closed under its binary operation as we remember from Chap. 5. What it meant was that whenever we multiply (or add) two elements of a group, the result is also in the same group.

Another part of the small spatial story of carrying objects around is that we can (almost) always put them back where they were. This is the sort of thing many of us are taught from the early childhood. You can play with your toys, but when you finish, please put them back where they were. And in the story of groups (and later rings and fields), it is called the inverse element. The group definition (an actor too, of course) demands that for every element in a group, there exists an inverse element.

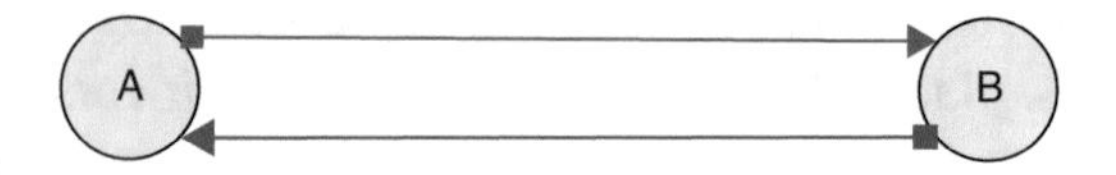

And when the inverse element is applied (added, carried, moved), the result is a no-move. And the no-move is also a vital character in the story of mathematical groups, where it is called "the identity element." The group definition, we discussed in Chap. 5, "insists categorically," as Herstein puts it, that such an element exists. A no-move or do-nothing element. Our experience of interacting with physical world tells us that when we do nothing to a stationary object, it stays where it is. In Newtonian mechanics it is referred to as "inertia" and is a consequence of the 1st law of Newtonian mechanics. If the net force acting on an object is zero, the velocity (including zero-velocity) stays constant. The inertia of physical objects, part of our everyday experience, is therefore, as we can see, a vital component of the story of mathematical groups, rings, fields, and—in a moment—also vector spaces. Adding zero, or performing a no-move, results in objects staying where they are. And exactly the same result can be achieved by moving an object and then putting it back where it was: $a + (-a) = 0$.

It is of course not for the first time that we encounter force-dynamic interactions as underpinning various elements of mathematical narrative. For example, in Chap. 3, we analyzed the story of mathematical proof, featuring an actor called the truth collector, who would move along a path, forcibly removing obstacles. In Chap. 5 we discussed group axioms as force-exerting actors, who would sometimes "fail to hold." But certainly, the force-dynamic base of algebra narrative is most clearly discernible in the story of mappings. And, as we have seen so far, this story is indeed one of the most often encountered building blocks of crucial definitions of algebra. The definitions of a group, a ring, a field, and (as we shall soon see) a vector space are all based on the concept of a mapping. All the binary operations (including the familiar addition, subtraction, multiplication, and division) are mappings.

And, while discussing Cayley's theorem in Chap. 5, we found that that the set on which group definition is based (the source story for groups) is a set of mappings (permutations). The binary operations in this set are therefore mappings of mappings. An ordered pair, one of the primary concepts from Chap. 4, is a mapping and order itself is a type of mapping too. Herstein's claim that "without exaggeration this is probably the single most important and universal notion that runs through all of mathematics" (1975: 10) is, as we have found out, perfectly accurate and not exaggerated indeed. And the first example of a mapping we encountered in Chap. 4 was $y = x^2$, "which **takes** every real number **onto** its square"[12] (ibid.).

[12] Emphasis added

Leonard Talmy (2000) observes that force dynamics seems to "underlie both our untutored 'commonsense' conceptions, and the sophisticated reasoning providing the basis for the scientific and mathematical tradition" (455). And, after our analysis of mathematical narrative, we can conclude that force-dynamic interactions underlie not only the process of reasoning (proof) but also its subject—the substance of mathematics.

> DEFINITION A nonempty set V is said to be a *vector space* over a field F if V is an abelian group under an operation which we denote by +, and if for every $a \in F$, $v \in V$, there is defined an element, written αv, in V subject to:
>
> 1. $\alpha(v + w) = \alpha v + \alpha w$.
> 2. $(\alpha + \beta)v = \alpha v + \beta v$
> 3. $\alpha(\beta v) = (\alpha\beta)v$.
> 4. $1v = v$.
>
> for all $\alpha, \beta \in F$, $v, w \in V$ (where the 1 represents the unit element of F under multiplication) (Herstein 1975: 171).

Further on, we learn that a module is a generalization of the concept of vector space in which the field F is replaced by a ring R.[13] A vector space is defined "over a field." As we remember from the previous chapter, a field is an abelian division ring. And in every ring, two binary operations are defined. So the definition of a vector space involves four binary operations: the multiplication and addition in F, the addition in V, and the multiplication of vectors by scalars, where vectors are in V and scalars are in F. The last of the four operations—the scalar-vector multiplication—is unusual in the sense that it is defined over two separate sets. It matches "interset" pairs (α, v) with elements of V. Or, to use another of the source stories for mappings we discussed in Chap. 3, it carries pairs (α, v) into V. We have four binary operations (mappings) and two sets of F and V featuring in the definition of a vector space. The binary operations (each of them a matchmaker) operate on Cartesian products of $F \times F$, $V \times V$, and $F \times V$, each of them requiring a separate "Cartesian" matchmaker to create. A vector space, we may conclude, requires the coordinated effort of seven matchmakers—we came a long way from a simple group, where just two of them were needed.

---

[13] The definition of a module is almost identical with the definition of a vector space, except for axiom 4, which is absent because a ring, as we remember, does not have to contain a "1" (a multiplication identity element).

Table 6.1 Elements of small spatial stories and traces of conceptual integration found in the narrative of rings, fields, and vector spaces

Objects	Elements of sets, numbers, vectors
Actors	Matchmakers (binary operations), carriers, axioms
Actions	Matching, carrying, moving, holding, resisting pressure
Image schemas	Container, in-out, source-path-goal, collection, link, object, process, compulsion, resistance, removing of restraint, counterforce, cycle, stacking up/removing (Jenga)
Conceptual blending	Various binary operations understood differently depending on which small spatial story is in the input space of the conceptual integration network. For example, in primary school arithmetic, the input for subtraction is always the "removing/taking away" small spatial story, which creates learning difficulties by making the odd numbers an odd concept for the young students

As we mentioned at the beginning of this section, vector spaces are indeed connected to geometry and physics. For example, all the equations of classical (Newtonian) mechanics feature R^3 vectors, where $R^3 = R \times R \times R$ is the triple Cartesian product of the set of real numbers, which is a vector space over the field of (again) real numbers.

Summary

In this chapter we analyzed the story of rings, fields, and vector spaces with reference (as before) to small spatial stories, image schemas, and conceptual blending. We learned on the way that the source story for rings are the familiar sets of integers, rational, and real numbers with their binary operations. The story of rings is also to be partially understood through the story of groups we analyzed in Chapter 5. Our findings are summarized in Table 6.1.

One of the crucial features of the algebra story so far was its firm rooting in the schemas of collection/container (sets and elements) and force/motion (mappings, axioms, proof). At one point we considered subtraction as adding an inverse element: $a + (-a) = a - a = 0$. And it seemed for a moment that the connection to the abovementioned schematic base was broken because schematically[14] $a + (-a) = 0$ means that two elements are added and as a result they both disappear. However, by returning for a moment to the chronicle of groups, we

[14] In the schema of adding objects to a collection or putting objects in a container.

were able to demonstrate that the schematic base consistently holds throughout the story so far. The narrative of algebra avoids the apparent problem of disappearing objects by using a different schema as a source for group binary operations. As we mentioned above, most crucial "abstract" mathematical notions are based on "concrete" entities. Groups, for example, have their origin in the set of permutations (one-to-one mappings of a set onto itself). And binary operations (such as addition) have their origin in compositions of permutations. The schematic equivalent of adding an inverse element is therefore (by strength of our analysis of mappings in Chap. 4 and Cayley's theorem) carrying a group of objects from place A to place B and then returning them to their original position. And thus the schematic base of the advanced algebra narrative is consistently preserved.

We also observed that this schematic consistence and simplicity of advanced algebra narrative does not find its counterpart in the way elementary arithmetic is taught at school. In a typical elementary arithmetic course, subtraction and the concept of zero are taught only after basic addition is mastered and are considered more difficult to grasp. Adding negatives comes later still and is considered to be even more daunting for students. Yet, as we have demonstrated by reading mathematical narrative closely, all three concepts of addition, subtraction, and zero are based on one schema only—moving objects from one place to another. And therefore all three are equivalent and complementary, which means they should not be taught separately. $a + (-a) = 0$ within this schema means moving objects and then returning them to the same place. As a result the objects stay where they were (the zero-move).

Vector fields with their intricate structure are as far as we travel in our exploration of modern algebra. At each point of our step-by-step linear analysis of the algebraic structures of growing complexity, we kept finding traces of conceptual integration, small spatial stories, and their building blocks—the image schemas.

Chapter 7
Summary and Conclusion

We proceeded as follows. After introducing the subject and method of research in Chap. 1, in Chap. 2 we presented the basic assumptions of the conceptual integration theory, with particular attention paid to small spatial stories and their basic ingredients—the image schemas. The small spatial stories always describe actors "moving and shaking" (changing location and manipulating objects). In the summary of each "research" chapter (Chaps. 3, 4, 5, and 6), we listed all the actors, objects, actions, and image schemas we managed to find, as well as "traces" of conceptual blending.

How did we find all of those? By reading Herstein's (1975) popular university-level algebra handbook. And not only the descriptive passages written in plain English but also the formulas like, for example, $A \cap (B \cup C) = (A \cap B) \cup (A \cap C)$ (cf. Sect. 3.9). The formulas are in fact also written in plain English, one just has to know how to read the symbols, and of course—this being a handbook—such instructions were provided. And this is how we know that the above formula reads "the union of set A and the intersection of sets B and C equals the union of the intersections of sets A and B and A and C." We followed all the crucial definitions (and the undefined "primitives"), theorems, and proof looking for image schemas, actors, actions, motion, space, objects, and traces of conceptual integration. For example, "$y = x^2$ [...] takes every real number onto its square" (Herstein 1975: 10, cf. Sect. 4.2) means that mapping is (can be conceived in terms of) an actor who carries objects (numbers) from one place to another. Or, more precisely, one of the inputs of the conceptual integration network

J. Woźny, *How We Understand Mathematics*, Mathematics in Mind,
https://doi.org/10.1007/978-3-319-77688-0_7

responsible for the meaning construction of mathematical mapping is the small spatial story of an actor carrying objects from one place to another. In Sect. 3.3.2, to give another example, we analyzed the definition of equal sets. "$A = B$, if both $A \subset B$ and $B \subset A$" (ibid.: 2), which reads "set A equals set B if both A is contained in B and B is contained in A." The meaning of the above can only be grasped if we can have access to two separate tokens (A and B) of a unique object (a blend of A and B). In other words, both inputs and the blend must be accessible for processing, which is consistent with the constitutive and governing principles of CIT, especially the "web principle" and the "unpacking principle" (cf. Sect. 2.2.4). The results of our search are gathered in Table 7.1.

In all the chapters, we noticed the crucial importance of selecting the correct small spatial stories and image schemas as inputs for the proper construction of mathematical meaning.

Knowing which ones to choose is a matter of success or failure for the students of mathematics. And the secrecy of this knowledge, only very obliquely hinted at in mathematical handbooks, creates part of the mystery, the aura of inaccessibility surrounding mathematics. For example, in Sect. 3.7 we found that although both sets and elements can be contained, different small spatial stories/image schemas apply in each case. Selecting the proper image schema/small spatial story also proved vitally important for understanding the homomorphism (cf. Sect. 5.9) or negative numbers and zero (cf. Sect. 6.7).

In Chap. 5, we talked about mathematical beauty to prove that it is accessible to ordinary mortals (i.e., non-mathematicians). And we were able to see it because all of us, despite the level of mathematical training, are endowed with both literary (Turner 1996) and mathematical mind. Mark Turner's "literary mind" is the mind which "works" like a literary parable, by projecting (blending) stories. As we tried to demonstrate by analyzing the language of modern algebra, the mathematical mind works in exactly the same way. The following quotation remains true when we exchange the adjective "literary" with "mathematical."

> The literary mind is not a separate kind of mind. It is our mind. The literary mind is the fundamental mind. [...] But the common view, firmly in place for two and a half millennia, sees the everyday mind as unliterary and the literary mind as optional. This book is an attempt to show how wrong the common

Table 7.1 Elements of small spatial stories and traces of conceptual integration found in the narrative of modern algebra

Objects	Elements of sets, numbers, all kinds of objects that can belong to a collection	Chap. 3
Actors	Set—an actor who possesses objects, governs property. Set operator (the potter, the setter)—performs operations on sets, uniting, intersecting, and dividing them. Proof—an actor who collects mathematical proofs but sometimes has to dispose of them to clear the path on her way to the QED spot	
Actions	Possessing/belonging (often categorized as a state, or a potential to act, perhaps not a prototypical action but of course, like with all linguistic taxonomies, the border between state and action is fuzzy), combining sets, forming them into new ones (uniting), intersecting, dividing, disposing of objects	
Image schemas	Containers with discrete and dimensionless, or voluminous objects (partly opened or tightly shut), an empty container (the null/empty set), part-whole, in-out, full-empty compulsion, blockage, removal of restraint, enablement, source-path-goal, object, superimposition	
Conceptual blending	The equality symbol "=" always involves a blend (triggers a conceptual integration network). Multiple tokens of an object are compressed into a unique object. Yet, because the projections are bi-directional and the network is maintained (according to the web principle and the unpacking principle, cf. Sect. 2.2.4), the object can be "one and many" at the same time	
Objects	Numbers, elements of sets	Chap. 4
Actors	The carrier, the matchmaker, the hiker	
Actions	Carrying, associating, moving from x to y	
Image schemas	Source-path-goal, compulsion, link, matching, superimposition, diversion, object, container, process	
Conceptual blending	Input spaces of the conceptual integration network can contain: Ordered pairs, points on the plane, carrying objects, associating objects, motion along a path. The circularity of the "rigorous" definition may reflect the circularity inherent in the process of blending. In any conceptual integration network, the mapping is bi-directional	

(continued)

Table 7.1 (continued)

Objects	Group elements, ordered pairs, ordered triples, bricks, kaleidoscope	Chap. 5
Actors	Various matchmakers creating the group structure, the truth-collector (proof), builders, kaleidoscope user	
Actions	Matching, mortaring bricks together, creating the Cartesian square and binary operation structure, collecting, exerting force, turning the kaleidoscope, recreating, preserving the structure, carrying, moving	
Image schemas	Container, source-path-goal, collection, link, object, process, compulsion, resistance	
Conceptual blending	"The one and many problem," the identity and inverse elements in a group—Multiple tokens of a unique object	
Objects	Elements of sets, numbers, vectors	Chap. 6
Actors	Matchmakers (binary operations), carriers, axioms	
Actions	Matching, carrying, moving, holding, resisting pressure	
Image schemas	Container, in-out, source-path-goal, collection, link, object, process, compulsion, resistance, removing of restraint, counterforce, cycle, stacking up/removing (Jenga blocks)	
Conceptual blending	Various binary operations understood differently depending on which small spatial story is in the input space of the conceptual integration network. For example, in primary school arithmetic, the input for subtraction is always the "removing/taking away" small spatial story, which creates learning difficulties by making the odd numbers an odd concept for the young students	

view is and to replace it with a view of the mind that is more scientific, more accurate, more inclusive, and more interesting, a view that no longer misrepresents everyday thought and action as divorced from the literary mind. (Turner 1996: v)

Our goal, expressed in Chap. 1, was to "prove that mathematics relies on the iterative use of basic mental operations of *story* and blending and demonstrate exactly how those two mental operations are responsible for the effectiveness and fecundity of mathematics" (cf. Sect. 1.2). So far, in this summary, we have discussed only the first part of it—the use of basic mental operations—but what about "the effectiveness and fecundity of mathematics"? How exactly did we account for the amazing adaptability of mathematics—its ability to reliably model the ever-changing world around us? There is an easy way out. We could let Mark Turner, for example, do our work and quote *The Origin Of Ideas: Blending, Creativity And The Human Spark* (2014), where the author argues convincingly that human creativity in any area, mathematics included, has its origin in conceptual blending.

> The claim of this book is that the human spark comes from our advanced ability to blend ideas to make new ideas. Blending is the origin of ideas. (Turner 2014: 9)

And we could finish now. But instead, let us go back for a moment to what we found by reading an excellent algebra handbook (Herstein 1975) closely. In Chap. 4, for example, we established that the official, "rigorous" definition of a mapping—"the single most important and universal notion that runs through all of mathematics" (Herstein 1975: 10)—is circular. The definition is circular because it is based on the undefined notion of an ordered pair, which is a mapping (cf. Sect. 4.4). We have also found that mathematicians go around this problem by prompting a different way of meaning construction for this crucial notion. We are encouraged to think of a mapping in terms of small spatial stories of "the carrier," "the hiker," or "the matchmaker," and this is how the circularity is avoided (cf. Sect. 4.7). As we explained in Sect. 2.2.2, "thinking in terms of" (understanding one story through another, the parable) means constructing a conceptual integration network. Mathematics avoids being barren (circular) by incorporating the structured and dynamic small spatial stories as inputs for conceptual blending. And in this way, the small spatial stories and blending account for the fecundity of mathematics, preventing it from being barren. The flexibility of mathematics, its ability to keep up with the fast-developing technology and natural sciences, stems from contextually motivated polysemy of the crucial mathematical terms—polysemy based on the choice from the inventory of "small spatial stories." What we have just said was put much better, 30 years ago, by George Lakoff:

> There is nothing easy or automatic or magical about the success of mathematics in empirical domains. It arises from [...] understanding of the phenomena

> in ordinary, everyday terms, which are then translated into corresponding mathematical terms. It is the human capacity to understand experience in terms of basic cognitive concepts that is at the heart of the success of mathematics. (1987: 364)

We wish now this was an algebra handbook so we could add QED.

We promised the reader not to skip ahead and we did not. We followed the structure of mathematical narrative from its foundations up, without jumping floors—from the simplest ("primitive") notions of a set and element to more complex concepts of a mapping, a group, a subgroup, a homomorphism, ring, field, and vector space. But we certainly did not cover the whole of modern algebra. We hope, however, that this book may be useful as a systematic sketch of the mathematical coastline, drawn from the vantage point of conceptual integration theory. Other travelers, and many of them will be needed, will have to fill in all the topographical details we missed.

Bibliography

Alexander, J. (2011). "Blending in mathematics". *Semiotica*, Issue 187. Pages 1–48.
Bernays, P. (1935). "Platonism in Mathematics". Lecture delivered June 18, 1934, in the cycle of *Conferences internationales des Sciences mathematiques* organized by the University of Geneva. Translated from French by C. D. Parsons. http://www.phil.cmu.edu/projects/bernays/Pdf/platonism.pdf, accessed 2017-11-07.
Brandt, L., & P. A. Brandt (2005). "Making sense of a blend. A cognitive-semiotic approach to metaphor". *Annual Review of Cognitive Linguistics*, Issue 3. Pages 216–249.
Brandt, L. (2010). *Language and enunciation - A cognitive inquiry with special focus on conceptual integration in semiotic meaning construction*. Doctoral dissertation, Aarhus Universitet.
Bache, C. (2005). "Constraining conceptual integration theory: Levels of blending and disintegration". *Journal of Pragmatics*, Issue 37. Pages 1615–1653.
Cayley, A. (1854). "On the theory of groups as depending on the symbolic equation $\theta^n=1$". *Philosophical Magazine*, Issue 7(42). Pages 40–47.
Coulson, S. (2000). *Semantic Leaps: Frame-Shifting and Conceptual Blending in Meaning Construction*. Cambridge: Cambridge University Press.
Coulson, S. & T. Oakley (eds.). (2000). "Special issue on conceptual blending":. *Cognitive Linguistics*, Issue 11(3/4). Pages 175–360.
Danesi, M. (2016). *Language and Mathematics: An Interdisciplinary Guide*. New York: Mouton de Gruyter.
Evans, V. & M. Green. (2006). *Cognitive Linguistics: An Introduction*. Edinburgh: Edinburgh University Press.
Fauconnier, G. ([1985] 1994) *Mental Spaces*. Cambridge: Cambridge University Press.
Fauconnier, G. & E. Sweetser (eds.). (1996). *Spaces, Worlds and Grammar*. Chicago: University of Chicago Press.
Fauconnier, G. (1997) *Mappings in Thought and Language*. Cambridge: Cambridge University Press.
Fauconnier, G. & M. Turner. (1998). "Conceptual integration networks". *Cognitive Science*, Issue 22(2). Pages 33–187.
Fauconnier, G. (1999). "Methods and generalizations". In T. Janssen & G. Redeker (eds.), *Cognitive Linguistics: Foundations, Scope, and Methodology*. Pages 98–128. Berlin, New York: Mouton de Gruyter.
Fauconnier, G. & M. Turner. (2002). *The Way We Think: Conceptual Blending And The Mind's Hidden Complexities*. New York: Basic Books.

J. Woźny, *How We Understand Mathematics*, Mathematics in Mind,
https://doi.org/10.1007/978-3-319-77688-0

Frege, G. (1879). "Frege (1879) Begriffsschrift, a formula language, modeled upon that of arithmetic, for pure thought" http://dec59.ruk.cuni.cz/~kolmanv/Begriffsschrift.pdf, accessed 2017-12-28.

Hausdorff, F. (1914). *Grundzüge der Mengenlehre.* Leipzig: Veit.

Gibbs, R. W. & G. Steen. (1999). *Metaphor in Cognitive Linguistics.* Amsterdam: John Benjamins.

Gibbs, R. W. (2000). "Making good psychology out of blending theory". *Cognitive Linguistics*, Issue 11(3/4). Pages 347–358.

Goldberg, A. (1995). *Constructions: A Construction Grammar Approach to Argument Structure.* Chicago: University of Chicago Press.

Harder, P. (2003). "Mental Spaces: Exactly when do we need them?". *Cognitive Linguistics*, Issue 14(1). Pages 91–96.

Harder, P. (2007). "Cognitive Linguistics and Philosophy". In D. Geeraerts & H. Cuyckens (eds.), *The Oxford Handbook of Cognitive Linguistics.* Pages 1241–1265. Oxford: Oxford University Press.

Herstein, I. (1975). *Topics in Algebra.* New York: John Wiley & Sons.

Hougaard, A. (2004). *"How're we doing?": An Interactional Approach to Cognitive Processes of Online Meaning Construction.* Doctoral dissertation, University of Southern Denmark, Odense.

Hougaard, A. (2005). "Conceptual disintegration and blending in interactional sequences: A discussion of new phenomena, processes vs. products, and methodology". *Journal of Pragmatics*, Issue 37. Pages 1653–1685.

Hohol, M. (2011). "Matematyczność ucieleśniona". In B. Brożek, J. Mączka, W.P. Grygiel, M. Hohol (eds.), *Oblicza racjonalności: Wokół myśli Michała Hellera.* Pages 143–166. Kraków: Copernicus Center Press.

Johnson, M. (1987). *The Body in the Mind.* Chicago: University of Chicago Press.

Koestler, A. (1964). *The Act of Creation.* New York: Macmillan.

Lakoff, G. & M. Johnson. (1980). *Metaphors We Live By.* Chicago: University of Chicago Press.

Lakoff, G. (1986). "A Figure of Thought". *Metaphor and Symbol*, Issue 1(3). Pages 215–225.

Lakoff, G. (1987). *Women, Fire and Dangerous Things. What categories reveal about the mind.* Chicago: Chicago University Press.

Lakoff, G. & M. Turner. (1989). *More Than Cool Reason: A Field Guide to Poetic Metaphor.* Chicago: University of Chicago Press.

Lakoff, G. (1990). "The invariance hypothesis: is abstract reason based on image schemas?", *Cognitive Linguistics*, Issue 1. Pages 39–74.

Lakoff, G. (1993). "The contemporary theory of metaphor", in A. Ortony (ed.), *Metaphor and Thought.* Pages 202–251. Cambridge: Cambridge University Press.

Lakoff, G. & M. Johnson. (1999). *Philosophy in the Flesh: The Embodied Mind and Its Challenge to Western Thought.* New York: Basic Books.

Lakoff, G. & R. Núñez. (2000). *Where Mathematics Comes From: How the Embodied Mind Brings Mathematics into Being.* New York: Basic Books.

Langacker, R. (1991). *Foundations of Cognitive Grammar, Volume II.* Stanford CA: Stanford University Press.

Leśniewski, S. (1913). "Krytyka filozoficznej zasady wyłączonego środka". *Przegląd Filozoficzny*, Issue 16. Pages 315–352.

Leśniewski, S. (1930). "O podstawach matematyki". *Przegląd Filozoficzny*, Issue 30. Pages 165–206.

Mac Lane, S. (1986). *Mathematics, Form and Function.* Berlin: Springer-Verlag.

Mandler, J. M. (1992). "How to Build a Baby: II. Conceptual Primitives". *Psychological Review*, Issue 99(4). Pages 587–604.

Mandler, J. & C. P. Canovas. (2014). "On defining image schemas". *Language and Cognition*, Issue 6(4). Pages 510–532.

Núñez, R. (2006). "Do Real Numbers Really Move?". In R. Hersh (ed.), *18 Unconventional Essays on the Nature of Mathematics.* Pages 160–181. New York: Springer.

Rohrer, T. (2005). “Mimesis, artistic inspiration and the blends we live by”. *Journal of Pragmatics*, Issue 37. Pages 1686–1716.

Rosch, E. H. (1978). “Principles of categorization”. In: E. Rosch & B. Lloyd (eds.), *Cognition and Categorization.* Pages 27–48. Hillsdale, N.J.: Erlbaum Associates.

Sinha, C. (1999). “Grounding, mapping, and acts of meaning”. In T. Janssen & G. Redeker (eds.), *Cognitive Linguistics: Foundations, Scope and Methodology*. Pages 223–255. Berlin: Mouton de Gruyter.

Stadelmann, V. (2012). *Language, cognition, interaction: Conceptual blending as discursive practice.* Doctoral dissertation. http://geb.uni-giessen.de/geb/volltexte/2012/8854/, accessed 2017-10-27.

Stockwell, P. (2002). *Cognitive Poetics: An Introduction*. London: Routledge.

Sweetser, E. (1990). *From Etymology to Pragmatics: Metaphorical and Cultural Aspects of Semantic Structure*. Cambridge: Cambridge University Press.

Talmy, Leonard. (1988). “Force Dynamics in Language and Cognition”. *Cognitive Science*, Issue 12. Pages 49–100.

Talmy, L. (2000). *Toward a Cognitive Semantics*. Cambridge: The MIT Press.

Turner, M. (1996). *The Literary Mind.* Oxford & New York: Oxford University Press.

Turner, M. (2005). “Mathematics and Narrative”. Paper presented at the International Conference on Mathematics and Narrative, Mykonos, Greece, 12-15 July 2005. http://thalesandfriends.org/wp-content/uploads/2012/03/turner_paper.pdf, accessed Nov. 11, 2016.

Turner, M. (2012). “Mental Packing and Unpacking in Mathematics”. In Mariana Bockarova, Marcel Danesi, and Rafael Núñez (eds.), *Semiotic and Cognitive Science Articles on the Nature of Mathematics*. Pages 248–267. Munich: Lincom Europa.

Turner, M. (2014). *The Origin Of Ideas: Blending, Creativity And The Human Spark.* Oxford & New York: Oxford University Press.

Van der Waerden, B. L., (1930). *Moderne Algebra*. Berlin: Springer.

Van de Walle, J. (2007). *Elementary and Middle School Mathematics Teaching Developmentally.* Boston: Allyn and Bacon (Pearson).

Wigner, E. (1960). “The Unreasonable Effectiveness of Mathematics in the Natural Sciences”. *Communications in Pure and Applied Mathematics*, Issue 13(I). Pages 1–14.

Printed in the United States
By Bookmasters